T0143305

Acceleration of Biomedical Image Processing with Dataflow on FPGAs

RIVER PUBLISHERS SERIES IN INFORMATION SCIENCE AND TECHNOLOGY
Volume 22

Series Editors

K. C. CHEN
National Taiwan University
Taipei, Taiwan

SANDEEP SHUKLA
Virginia Tech
USA

CHRISTOPHE BOBDA
University of Arkansas
USA

The "River Publishers Series in Information Science and Technology" covers research which ushers the 21st Century into an Internet and multimedia era. Multimedia means the theory and application of filtering, coding, estimating, analyzing, detecting and recognizing, synthesizing, classifying, recording, and reproducing signals by digital and/or analog devices or techniques, while the scope of "signal" includes audio, video, speech, image, musical, multimedia, data/content, geophysical, sonar/radar, bio/medical, sensation, etc. Networking suggests transportation of such multimedia contents among nodes in communication and/or computer networks, to facilitate the ultimate Internet.

Theory, technologies, protocols and standards, applications/services, practice and implementation of wired/wireless networking are all within the scope of this series. Based on network and communication science, we further extend the scope for 21st Century life through the knowledge in robotics, machine learning, embedded systems, cognitive science, pattern recognition, quantum/biological/molecular computation and information processing, biology, ecology, social science and economics, user behaviors and interface, and applications to health and society advance.

Books published in the series include research monographs, edited volumes, handbooks and textbooks. The books provide professionals, researchers, educators, and advanced students in the field with an invaluable insight into the latest research and developments.

Topics covered in the series include, but are by no means restricted to the following:

- Communication/Computer Networking Technologies and Applications
- Queuing Theory
- Optimization
- Operation Research
- Stochastic Processes
- Information Theory
- Multimedia/Speech/Video Processing
- Computation and Information Processing
- Machine Intelligence
- Cognitive Science and Brian Science
- Embedded Systems
- Computer Architectures
- Reconfigurable Computing
- Cyber Security

For a list of other books in this series, visit www.riverpublishers.com

Acceleration of Biomedical Image Processing with Dataflow on FPGAs

Frederik Grüll

Goethe University Frankfurt,
Germany

Udo Kebschull

Goethe University Frankfurt,
Germany

River Publishers

Published, sold and distributed by:
River Publishers
Alsbjergvej 10
9260 Gistrup
Denmark

River Publishers
Lange Geer 44
2611 PW Delft
The Netherlands

Tel.: +45369953197
www.riverpublishers.com

ISBN: 978-87-93379-36-7 (Hardback)
      978-87-93379-35-0 (Ebook)

©2016 River Publishers

Contents

Foreword

The success of static Multiscale Dataflow Computing is an acknowledgement of the rising importance of Data as opposed to the algorithms. As transistors shrink, wire delays become more important than logic delays and memory accesses matter more than arithmetic operations. Therefore, in this new world, an optimal computing methodology is one that minimizes the total travel distance of all data items while an algorithm is running. To minimize the sum of all distances that the data needs to travel, we need to organize the arithmetic units to put data producers and data consumers as close together as possible. Luckily, this seemingly difficult problem can be solved optimally by tools working together with the dataflow programmer, or as Daniel Slotnick, the chief architect of the ILLIAC IV said: "The parallel approach to computing does require that some original thinking be done about numerical analysis and data management in order to secure efficient use."

The authors of this book are skillfully demonstrating the efficiency of the dataflow paradigm and the easy-of-use of the tool-chain, generating a practical hardware-optimal solution by optimizing their implementation on multiple levels of abstraction (and multiple scales of granularity) or as scientists call it, Multiscale optimization.

Now, looking at the big picture, why would a new computing paradigm based on dataflow (executing graphs) have a chance to change the world in 2016? A possible reason is The Cloud! As it turns out, in The Cloud, suddenly new hardware paradigms can be deployed quickly and at scale as long as the new paradigm brings either cost reduction ("in the Cloud, cost is everything") or differentiation (or both). As a side effect, in a Cloud-driven world, innovation moves from established system integrators and chip manufacturers to whoever controls the Cloud. Learning from Dan Slotnick's words from over half a century ago, we know that the parallel approach requires effort, and after many years of adding overhead to make programming easier for the masses, there is a key opportunity for Multiscale Dataflow Computing to restore efficiency and at the same time make computing more predictable.

What are the consequences of Multiscale Dataflow Computing on Computer Science? Does dataflow computing follow classical Complexity Theory or do we need a new, communication-centric Complexity Theory? If the complexity of a computation lies in the shape and size of it's data rather than operations, how does one describe for which data set it is more efficient to compute using a dataflow paradigm, and for which data set it is more efficient to use a CPU, a pocket calculator or just do the math in one's head. Does it make sense that no single algorithm can be "suitable" or "not-suitable" for Multiscale Dataflow Computing, but suitability solely and counter-intuitively, rests in the complexity of the problem-specific data? How can we create a new concept of Data Complexity, which of course needs to be described in conjunction with an algorithm's data access pattern, and the algorithm's interaction with the memory hierarchy of a computer system. Maybe we could call the resulting architecture a *communication-avoiding architecture*?

How does Multiscale Dataflow Computing with the concept of Data Complexity, make computing more predictable? The same way that a factory makes production and consumption of goods predictable, a static Multiscale Dataflow Computer makes it easy for us to calculate performance and evaluate different algorithmic options without having to implement them first. Maybe now that CPU clock frequencies are not going up any more, Multiscale Dataflow Computing can help to give us the next orders of magnitude of performance improvement to bring advances to our planet which we currently could not even dream of.

Oskar Mencer

Preface

Computing has seen a major shift on all levels since the beginning of the century. Before, software applications were expected to scale, but it was also a given as granted that single-core processor performance speeds would continue to increase exponentially. Software that did miss a performance target by a constant factor would meet it soon with the next hardware generation. This almost effort-less acceleration came to a halt when clock frequencies of CPUs could not be increased any further due to uncontrollable heat dissipation above 4 GHz. Since then, acceleration needs to be sustained by different methods, and computing has been focused on parallel execution for all application areas with a need for performance.

In this book we want to introduce dataflow computing on FPGAs as an alternative to multi-core computing, vector instructions and graphics cards for image processing. Instead of computing the same instruction for multiple data items, dataflow computing processes an algorithm in a system of deep pipelines, where the data moves through one step at a time. Image processing is well-suited for this kind of computing because many algorithms can be re-written such that a pipeline processes one pixel per clock cycle. At the moment, however, there is few literature available that covers program acceleration with dataflow computing from end to end. We hope to close this gap with this book.

The book is aimed towards students and professionals who want to learn about this execution model and make it usable for high-performance image processing. In order to create a book usable for developers faced with a performance problem and learners alike, we decided to follow the process of accelerating an algorithm, starting with an introduction covering the basics of dataflow and finishing with a working accelerator. This lead us to discuss the identification of bottlenecks, the transformation of algorithms from an imperative description towards a dataflow graph and last, but not least, the preservation of accuracy.

The compiler and dataflow engine of Maxeler Technologies were used to program the example applications. A free version of the compiler for hardware and simulation targets is available without charge after registration at the Maxeler University program MAX-UP. We aimed to keep the book as agnostic as possible with regard to the chosen toolchain and FPGA hardware and hope that it can also be of good value for readers that program in other high-level languages with dataflow support.

Acknowledgements

This book could not have been written without the support of our colleagues and collaborators. We especially like to thank Manfred Kirchgessner for his work on feature extraction for Localization Microscopy during his final year thesis. We are also very grateful for discussions about hardware synthesis that we had with Heiko Engel, Sebastian Manz and Torsten Alt.

The groups of Prof. Michael Hausmann and Prof. Christoph Cremer at Heidelberg University provided localization imagery that allowed us to test our algorithms and develop them towards robust tools that are safe to operate in research labs. The research group of Prof. Achilleas Frangakis supported us with algorithms for electron tomography. The employees of Maxeler Technologies helped us to understand dataflow computing and apply it to image processing as members of the Maxeler University program MAX-UP. These contributions were critical for writing a book about dataflow computing in practise.

Research for this book was supported in part with funds from the Frontier innovation fund of the University of Heidelberg and the Helmholtz Graduate School for Hadron and Ion Research.

List of Figures

List of Tables

List of Listings

List of Abbreviations

AD	anno Domini
ALU	arithmetic logic unit
AMD	Advanced Micro Devices, Inc.
AP	access pattern
ART	Algebraic Reconstruction Technique
ASIC	application specific integrated circuit
ASC	A Stream Compiler for Computing with FPGAs
BRAM	block random-access memory
CCD	charge-coupled device
CE	clock enable
CLB	configurable logic block
CPU	central processing unit
CSP	communicating sequential processes
CSV	comma-separated values
CT	computed tomography
CUDA	compute unified device architecture
DAG	directed acyclic graph
DDR	double data rate
DFE	dataflow engine
DIP	digital image processing
DMA	direct memory access
DRAM	dynamic random-access memory
DSP	digital signal processing
FF	flip-flop
FIFO	first in, first out
FIR	finite impulse response
FP	floating point
FPGA	field programmable gate array
FSM	finite-state machine
GB	gigabyte
GHz	gigahertz

GPS	Global Positioning System
GPU	graphics processing unit
GUI	graphical user interface
HLS	high-level synthesis
IC	integrated circuit
IDE	integrated development environment
IEEE	Institute of Electrical and Electronics Engineers
IO	input/output
IOB	input/output buffer
IP	internet protocol
IPC	instructions per clock cycle
ISBI	International Symposium on Biomedical Imaging
ISE	integrated software environment
JPEG	Joint Photographic Experts Group
KB	kilobyte
KDF	Khoros Standard Data Format
LSB	least significant bit
LUT	lookup table
MB	megabyte
MHz	megahertz
MIMD	multiple instruction streams, multiple data streams
MISD	multiple instruction streams, single data stream
MIT	Massachusetts Institute of Technology
MPx	megapixel
MPEG	Moving Picture Experts Group
MPI	Message Passing Interface
OSC	oscillator
PATTSY	Processor Array Tagged-Token System
PCIe	Peripheral Component Interconnect Express
PDF	Portable Document Format
PMU	performance monitoring unit
POSIX	Portable Operating System Interface
RAM	random-access memory
RISC	reduced instruction set computing
ROB	re-order buffer
ROCCC	Riverside Optimizing Compiler for Configurable Computing
ROI	region of interest

ROM	read-only memory
RTL	register-transfer level
SART	Simultaneous Algebraic Reconstruction Technique
SD	standard deviation
SIMD	single instruction stream, multiple data streams
SISAL	Stream and Iterations in a Single Assignment Language
SISD	single instruction stream, single data stream
SNR	signal-to-noise ratio
SO-DIMM	small outline dual in-line memory module
SPDM	Spectral Precision Distance Microscopy
SPIDER	System for Processing of Image Data from Electron microscopy and Related fields
SRAM	static random-access memory
SSA	static single assignment
SSE	streaming SIMD extensions
STED	stimulated emission depletion
STORM	stochastic optical reconstruction microscopy
TIFF	Tagged Image File Format
USA	United States of America
VAL	Value-Oriented Algorithmic Language
VHDL	VHSIC Hardware Description Language
VHSIC	Very High Speed Integrated Circuit
Xmipp	X-Window-based Microscopy Image Processing Package

1

Introduction

1.1 Motivation

Computing has seen a major shift on all levels since the beginning of the century. Before, software applications were expected to scale, but it was also given as granted that single-core processor speeds would continue to increase exponentially. Software that did miss a performance target by a constant factor would meet it soon with the next generation of hardware. This almost effortless acceleration came to a halt when clock frequencies of CPUs could not be increased any more due to uncontrollable heat dissipation above about 4 GHz. Since then, acceleration needed to be sustained by a different method, and computer programming increasingly focused on parallel computing. Here, the computer hardware executes multiple instructions at the same time, and the total throughput is increased by the parallel factor in the best case. CPUs have gained the ability for parallel execution with multi-core processors, and vector instructions were rediscovered to speed up execution on the instruction level. Modern CPU architectures can still be programmed in a sequential manner, but the most compute-intense parts are now routinely converted to multi-threaded execution or vector instructions. All major computer languages have shifted to offer support for parallel programming, by either providing library support or sometimes through language extensions.

A massive parallelism suitable for practical use was discovered in about 2005 when graphics cards proved to be capable as application accelerators outside of the realm of computer graphics. On these cards, hundreds of compute cores that resemble an array of simple processors were able to accelerate certain computing problems by more than two magnitudes compared to CPUs. Graphics cards quickly gained popularity in high-performance computing, despite their intrinsic execution model that restricts acceleration to massively parallel programs where each thread group performs the same step at a time.

An even finer grained execution model can be found in field programmable gate arrays (FPGAs), the hardware used in this book. FPGAs can be programmed below the level of the processing unit. A large array of very basic hardware elements can be configured to emulate every other digital computer chip very close to its hardware level, given the FPGA does not run out of resources. The increasing size of FPGAs has moved their purpose from small logic applications toward accelerators for fully featured applications in high-performance computing. The tremendous flexibility of their internal and peripheral architecture promises to accelerate all sorts of computing problems with less restriction that graphics cards impose on the algorithms of the application.

FPGAs can be found in areas where high computing performance or low latency is such a hard requirement that custom computing becomes feasible, but application-specific integrated circuits (ASICs) remain too expensive due to low purchasing volumes. They are used in experiments for high-energy physics where trigger latencies must low, in electronic high-frequency trading or seismic exploration. The acceleration gained in these areas motivates the extension of the scope of application toward other high-performance applications in science, such as image processing and reconstruction.

In this book, the usage of FPGAs as application accelerators is examined for the domain of biomedical image processing and reconstruction. Here, many compute-intense problems can be found that have not yet attracted the interest of commercial vendors in particular, and where the increased size of modern FPGAs and new programming tools for dataflow computing tools have only recently made it feasible to port them from CPUs or graphics cards toward dataflow engines that build on FPGAs.

Shortened computing times are crucial for timely diagnostics. Data analysis that is performed in real time with data acquisition moves diagnostics from a work flow interruption toward an interactive tool. Samples could be examined second time before they decay, diagnosis would become faster, and laboratory examination has the potential to become cheaper and therefore promotes its usage.

1.2 Overview

1.2.1 The Idea

An algorithm that was written for high-performance computing on a CPU or graphics card reflects many optimizations that were chosen on purpose during performance evaluation, but also implicitly based on past experience of the programmer. FPGAs can emulate CPUs, but only slower by at least one order of magnitude. For efficient execution that achieves an acceleration over the previous hardware architecture, an FPGAs application must be dissected and re-implemented such that the unique architecture is employed not for emulation, but for custom-designed computing. The most efficient implementation for FPGA computing is usually the pipeline, a system of hardware operators and wired connections in the FPGA where every operator performs one operation in every time step. Laid out with a length of hundreds of pipeline stages, a result is produced at its output with every cycle in the best case, and since every operator executes all the time, the design is optimal for algorithms that can be molded into such a structure.

To reduce the cognitive load of not only understanding the algorithm of the program to port from the source code, but also the description of the hardware, it was chosen to use a so-called dataflow compiler from Maxeler Technologies for pipeline description. FPGA programming is traditionally performed in languages that describe the behavior on a very low level, and hardware programming is observed to consume much more time for the same functionality than software programming. Because the features aimed for data manipulation at the lowest level of traditional hardware programming languages were not needed, a higher level compiler for pipeline description could be used to bring down the development time and target the acceleration of two applications with a FPGA-based dataflow engine within this book.

1.2.2 Aim of this Book

Two applications were suffering from slow execution times and were selected as acceleration opportunities to develop a guide about how to port software written in an imperative language toward a dataflow description. From the dataflow description, they can then be automatically translated to a system of pipelines on an FPGA. The observations and insights gained during the porting process form the center of this book. Most of the procedures that help to bridge the gap have been used before, but were not collected and written down in a concise manner. The second aim was to actually accelerate the applications and provide value for its users in terms of runtime and data throughput. For this, other factors such as maintaining the accuracy of the results and usability became important.

The example applications cover common algorithms which can be found in most related applications in image processing. The first application is used for analyzing raw data from localization microscopy. By marking important molecules in (live) biological structures and switching them stochastically between a bright and a dark state under a conventional light microscope, localization microscopy can separate the blurry optical signals in the time domain in a second step on a computer and eventually produce an image with a resolution improved by one order of magnitude. For this application, a CPU implementation was available and needed to be ported in a combination of algorithmic and hardware design.

The second application reconstructs 3D images from 2D projections for electron tomography, a special case of computed tomography. Due to imprecision in the alignment of the electron microscope, the problem cannot be reduced to a set of 2D reconstructions, and hence, the algorithms developed for 2D tomography are unsuitable. Electron tomography provides a much higher resolution, but requires the sample to be dead. It complements optical light microscopy on the cellular level and generates 3D images. The source code was available in the CUDA language for graphics cards.

We want to show that dataflow computing can be used today for efficient acceleration of time-consuming computing problems. In particular, the following items that have been considered the main obstacles for dataflow computing can be overcome for image processing.

1. **Effective portability from an imperative program to a dataflow description**
   It is shown that concepts from applications used for image processing and reconstruction can be ported to a dataflow description that is suitable for FPGAs. In a continuation and as advancement to previous work, compute-intense application kernels can now be ported to FPGAs that were excluded before by the limited size of previous FPGAs and programming tools.

2. **Efficient development with high-level languages**
   Custom hardware is still mostly developed with low-level languages, either VHDL or Verilog. It is shown that a description on a higher level with dataflow programming as the fundamental programming paradigm increases development efficiency and enables complex designs that would not be possible with the legacy programming tools within the given time.

3. **Acceleration of both sample applications by a significant factor**
   An acceleration was achieved for both the applications that were implemented. For image analysis in localization microscopy, a total acceleration of a factor of 18,500 was achieved, with about equal parts

owed to algorithmic redesign and hardware acceleration. The application for electron tomography was accelerated by a factor of five compared to an implementation on graphics cards.

1.3 Outline

After this introduction, the basic principles of dataflow computing and FPGA-based dataflow engines (DFEs) are introduced and their usage for application acceleration. Dataflow computing is presented as an efficient design method for high-performance computing with a focus on image processing. An overview about the state of the art of the still new concept of exploiting reconfigurable computing for data processing is given. As the hardware continues to get more capable with every generation, the tools to describe problems for application accelerators must keep pace and are covered in the second part of Chapter 2.

The translation process from imperative languages, such as C, requires rethinking of the algorithms of a program. A general top–down approach for an efficient porting process is presented in Chapter 3, "Acceleration of Imperative Code with Dataflow Computing." After the computational bottlenecks of a program are understood, the chapter describes how to pipeline a previously sequential control flow. The chapter concludes with hardware details that determine the efficiency of logic operations and memory access.

This top–down approach is then used to accelerate two example applications in Chapter 4. Localization microscopy and electron tomography both require high-performance processing of image data and were chosen to benchmark the DFE as well as the approach of combining algorithmic with hardware design.

This book is completed with the conclusion in Chapter 5. Dataflow computing is confirmed to be well-suited for algorithms in image processing that were previously too big or too complicated for previous FPGA generations and for the constraints of legacy programming tools.

2

Dataflow Computing

2.1 Early Approaches

The first dataflow machines were devised in the late 1970s [2]. They provided an alternative to the von-Neumann architecture that forms the base of general-purpose computing until today. In a von-Neumann machine, one instruction is executed at a time, and instructions and data are stored in the same memory. Modern CPUs deviate from this method to a certain extend with the integration of multiple cores that compute independently, with reordering of instructions to account for data dependencies and with vector units that perform the same instructions on multiple data items. The programming model in high-performance computing, however, did not change much. Each thread of execution is still described as a series of instructions in imperative languages like C and Fortran, and the program is close to the linear execution model of a CPU core. In this section, the dataflow model of computation is introduced, and its implications to parallel computing are discussed.

2.1.1 Control Flow and Dataflow

A program written in an imperative language consists of segments of sequential instructions that manipulate data. Starting with the first instruction, each segment describes an algorithm that is executed linearly until the end is reached. These linear parts are controlled by control flow statements to choose which part to execute, depending on input data or intermediate results. Control statements can be conditionals that select one of multiple sequences for execution, or loops that control the repeated execution of a sequence. *If-else* and *select* statements are popular choices in language design for conditionals and loops are usually implemented as *while, do-while, for-loops,* or sometimes tail-calls in recursion. These statements define the control flow explicitly, and the programmer chooses the branching or looping condition depending on the dataflow of the algorithm.

7

The dataflow of an algorithm is defined by the data that are exchanged between instructions. The output of one instruction may be the input of one or more following instructions, which will in turn produce more data that are used in later stages when their operations are executed. Hence, the dataflow of a program pictures the flow of data between operations, and it is the task of the programmer in an imperative language to use conditionals and loops to ensure that all data are processed in the right order and that every instruction is executed only if all its input data are valid and available.

Imperative programs for high-performance computing hit a limit in throughput when the clock frequency of the CPU does not allow for a faster scheduling of instruction execution. Instead, the program has to be executed in parallel, with multiple instructions being executed at the same time. Only few hardware improvements can help achieve parallelism in a CPU without requiring effort from the programmer, such as superscalar architectures that execute more than one instruction in parallel. Most other techniques force the programmer to think about the dataflow and write down which parts of the program can be run in parallel through vector instructions, hints to the compiler or explicit multi-threading with locking to preserve data consistency. Multi-threading an application is the only way to use multi-core CPUs efficiently. It is notoriously hard for nontrivial algorithms to achieve without the introduction of data races or deadlocks [3].

Dataflow programming aims to implement an algorithm by keeping the control flow of a program implicit. Instead, the dataflow between operations is described. Dataflow descriptions resemble a graph where a node is drawn for every operation and a directed edge for every forward of a result from one operation to an input of another operation. This dataflow graph then describes the relations between all operators. Besides this, it can also be used to determine which parts of the algorithm can be run in parallel on a very fine-grained level: Each operator that has valid data on all of its input edges and is not stalled at its output is free to run. The scheduling is achieved locally for every operator just by observing its inputs and output, and the system is capable to operate on the maximum degree of parallelization possible.

An example dataflow graph can be seen in Figure 2.1. It computes the center of mass of an input stream with weights q. The positions $x = 0, 1, 2, \ldots$ are generated from a counter. With every cycle, a data item travels along the edges and gives the center of mass of the weights processed so far at the output. This logic could be used for feature extraction from a line of pixels in an image. When the input does not run empty and the output does

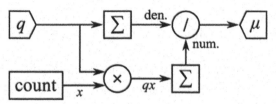

Figure 2.1 Dataflow graph of a center-of-mass calculation. The directed graph features an input (q), a generator (count), two stream operators ($\times, /$), two accumulating reductions ($\sum$), and an output (μ). It describes the calculation of $\mu = \sum_i q_i x_i / \sum_j q_j$ for all weights q that were supplied at its input. The values of x are generated by the counter.

not stall, all operations can run in parallel and the center of mass is updated with every cycle at the output of the graph.

2.1.2 Dataflow Machines

Dataflow programs can be implemented in software for CPUs by storing the dataflow graphs in the main memory and checking each node for valid inputs and outputs repeatedly. The instruction of a node is then executed if possible to create new data that propagate further through the graph. This approach suffers from a large overhead caused by the repeated checking of each node. To increase the performance, dataflow machines were developed that accelerated the scheduling of each node with dedicated hardware.

In most imperative languages, a function can be entered more than once at a time. This allows a function to call itself directly or through other functions and enables recursion. The context of each call is stored separately in frames on the stack, and each stack frame contains the values of the local variables. An implementation of a dataflow graph like the one shown in Figure 2.1 cannot run multiple instances of the same graph concurrently as it would be needed to support recursion. Instead, when run with data that need to be kept separate, the accumulators would still add all data independent of their context and update only one result in the internal state of the node. To support recursion, it is therefore needed to "tag" the data with its context and implement the nodes of dataflow graph such that they only operated on data with the same tag when executed. These data items are also referred to as "tokens."

The support or lack of for recursion is the most defining characteristic of dataflow machines. Static systems do not separate data travelling along the nodes of the graph and can therefore not support reentrant execution of subgraphs needed for recursion. Instead, the subgraph must be instantiated multiple times in memory. Dynamic systems use tagging that allows concurrent execution of a subgraph and support recursion. Tagging can also help to perform different iterations in a dataflow graphs with loops more efficiently.

For static systems, one of the earliest examples is the MIT static dataflow machine prototype [4]. Its design is shown in Figure 2.2. The machine stores a representation of the dataflow graph in its random access memory and assigns three data words to each node. The first word is read-only during execution and defines the kind of operation to be performed by the node. The other data words contain the input values of the operation if present. Operations that have more than two inputs need to be constructed with multiple binary nodes. If the machine detects two valid input words for a node, an arbitration network sends all three values to a processor unit called instruction cell where the operation is executed. The result is send back and forms an input token for one or multiple nodes until an exit node is reached and an end result can be produced. Other implementations of static dataflow machines are the NEC image pipelined processor [5] and Hughes dataflow multiprocessor [6].

Dynamic dataflow machines that contain a tag for each data token increase the performance by supporting the processing of data from different loop iterations or function calls in the graph at the same time and can therefore provide a better utilization by increasing parallelism. An example is the Manchester dataflow computer [7]. It features a circular design similar to the static dataflow machines: Tokens are collected in a matching unit, where nodes with valid inputs are detected. The input data are then send to one of multiple processing units where the instruction of the node is executed. Finally, the result is send back to a FIFO leading to the matching unit again.

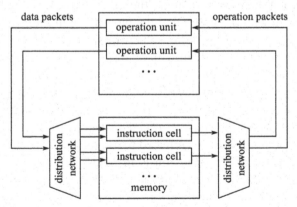

Figure 2.2 Design of the MIT static dataflow machine. The dataflow graph is stored in memory. Nodes that have valid input are sent to a processing unit for execution, and the result is fed back into the graph [4].

Contrary to static dataflow machines, the matching unit not only monitors for nodes that have input present, but also ensures that a node has a set of input tokens with the same tag. Tags are generated by special nodes at the entrance of loops or functions. Machines with a similar architecture are the MIT tagged-token dataflow machine [8] and the processor array tagged-token system (PATTSY) [9].

The level of parallelism is limited by the number of instruction cells and the ability to monitor the input of all nodes that reside in memory for scheduling. For systems with many instruction cells, the random memory access of node registers turned out to be a major bottleneck. A solution is to group nodes into supernodes to reduce the pressure through monitoring [2]. Inside these nodes, the execution follows a linear sequence of instructions and instructions follow a fully static scheduling.

The complex scheduling for both static and dynamic dataflow machines is caused by the desire to support "bubbles" between data tokens in the dataflow graph. These parts of the pipeline system where no data are present increase the flexibility of the system, but also increase the overhead because every node must observe to schedule its execution. Statically scheduled pipelines remove this overhead at the expense of flexibility. Here, all data move in lockstep through the graph, and every node is guaranteed to have valid inputs and outputs. The scheduling of the system can then be simplified to only observer the global inputs and outputs of the graph, and the overhead is reduced to enable the advancement of data on either all nodes or none. This type of dataflow machine became only possible with the advent of reconfigurable hardware. Its most common implementation, the field programmable gate array (FPGA), allows the hardware to be modified such that it represents the dataflow graph in space on the silicon, and all nodes can be executed at the same time with massive parallelism. FPGAs will be presented later in this chapter, followed by Chapter 3 where the implications of computing with statically scheduled pipelines and their performance benefit are presented.

2.1.3 Dataflow Programs

The graph of a dataflow program can be described graphically as in Figure 2.1 or with text. The graphical representation is usually used for visualization only. One of the first and most influential textual programming languages for dataflow computing is "A Value-Oriented Algorithmic Language" (VAL) [10]. An example is shown in Listing 2.1. A value can only be assigned once to an identifier during declaration. After that, the identifier holds the value and acts as a constant. The technique is called single assignment and leads to a close

```
function Stats (X, Y, Z: real returns real, real)
  let
   Mean real := (X + Y + Z) / 3;
   SD real := SQRT((X*X + Y*Y + Z*Z) / 3 - Mean
           * Mean);
  in
   Mean, SD
  endlet
endfun
```

Listing 2.1 A short program in VAL that describes the computation of the mean Mean and standard deviation SD of three input values X, Y, and Z [1].

relationship between code and graph. The assignments correspond to the result of a node in the dataflow graph and arithmetic operations on a value form the nodes. From the textual description, the compiler can interfere that the calculation of Mean and S.D. can be performed in parallel. Even inside of an expression parts can be computed in parallel, such as the products for the value of S.D. The actually achieved level of parallelism is eventually determined by the capabilities of the hardware.

VAL supports function calls, such as the call to the SQRT() function. The calls are inlined to produce the final dataflow graph; thus, recursion is not supported. Loops are restricted to linear iteration for mapping a list to a new one or for reducing to a scalar value. The program can run on both static and dynamic dataflow machines.

The "Stream and Iterations in a Single Assignment Language" (SISAL) provides a richer set of instructions to the programmer. It was derived from VAL, adds support for recursion and finite streams of values. It also eases the restriction of single assignment by supporting the assignment of a value to a new identifier with the same name [11]. Execution of SISAL programs requires a dynamic dataflow machine, such as the Manchester dataflow machine. As an alternative, SISAL programs can be translated to C by the compiler and executed on most Unix systems.

2.2 Principles of Dataflow Computing on Reconfigurable Hardware

Dataflow machines consist of a fixed number of processing elements and a scheduler that arbitrates which node operation can be processed first. Reconfigurable hardware operates differently: Instead of utilizing the existing processing elements as well as possible, the hardware itself is modified to follow the dataflow graph. The dataflow graph is implemented in space with dedicated wiring for the edges of the graph and operations implemented in hardware for the each node. Before the reconfigurable hardware is introduced in the next section, an overview is given of the primitives of a dataflow graph that are close enough to be directly implemented in hardware.

2.2.1 Primitives

On the register-transfer level, a hardware design consists of registers and combinatorial logic. The primitives of dataflow computing, however, reside above the register-transfer level. They form a higher abstraction that covers arithmetic operators and storage.

Stream operators: Nodes with inputs and outputs that perform an operation on the inputs and forward the result to the output are named stream operators. Most of the actual computation is carried out by these operators that perform arithmetic or logic operations, select between sources as multiplexers or cast numbers between different formats, ranges, and precisions. Albeit stream operators may contain pipeline registers, they are stateless in a certain sense: The result of an operation does not affect the result of the following ones. For more complicated operations, such as floating-point processing, an operator may contain many stages of registers that increase its latency, but make high clock frequencies possible and therefore lead to a high throughput.

Generators: A node with no input can still be of use if it generates a stream from a constant value for a following calculation or a pattern of output values from an internal state. A counter, for example, may increase its internal value with every clock cycle by a fixed amount. The output can then be used to keep track of the position in the stream and perform special actions for boundary conditions, such as the border of rasterized image. Other use cases are generators for bit masks and random number generators.

Reductions: A stream reduction is a node that collects information in an internal state and modifies it depending on the input values. The canonical example is an accumulator that adds every input value on top of its inner state until it is reset. Other reductions accumulate bits in a register or track the greatest value in a stream that has passed through the node. The internal state is accessible at its output and often of interest only at the end of the calculation. They can usually be found at the tail of a pipeline.

Stream navigation: For some applications, it is required to compare a value in a pipeline with the previous one and calculate the difference. On the register-transfer level, this can be done with an extra registers that delay the current signal by one clock cycle. The delayed signal will then be provided at the output of the register. For pipelines that are scheduled automatically, a stream navigation node indicates the scheduler that a value is to be delayed. With more registers, the navigation range past the current value can be further extended, and a BRAM FIFO will allow to freely tap the data stream within the FIFOs range.

Inputs and outputs: The global inputs and outputs of the dataflow program are described with corresponding input and output nodes in the dataflow graph. Depending on the implementation, these nodes may often require the most resources of all nodes in order to receive and send data over the PCIe bus, the IP network, a video link, or another high-level protocol.

Memory: If reductions and stream navigation are not sufficient to maintain state information between items in the data stream, memory nodes provide storage with random addressing. On the FPGA, they are most likely provided by BRAM that can be read and written at the same time at every clock cycle, allowing one stream of values to be stored and one stream to be extracted in parallel. Alternatively, BRAM can be used as ROM with its content to be set during configuration of the FPGA. The addresses for read and write access are set explicitly from a second set of user-defined data streams.

External memory can be used when more than a few megabytes of data must be accessed. Usually implemented as dynamic RAM (DRAM), its latency is higher. Its relative cheap price makes it suitable to store many gigabytes of data, similar to the working memory of a CPU system.

Finite state machines: For more complicated state transitions than the ones generators or reductions offer, a programmer may describe a finite state machine (FSM) and embed it into the dataflow graph as a node. The state transition will then be executed either with every clock cycle for maximum flexibility, or when the pipeline advances for embedded FSMs. The FSM may be described in its own language such as an RTL language. FSMs are closer to von-Neumann machines by design and may not be supported by the dataflow description tool.

The dataflow developer has multiple options to describe the graph. Depending on the tool, it can be drawn in a graphical user interface, created as the result of a program with library support, or from a compiler that comes with its own descriptive language.

2.2.2 Scheduling

After the dataflow graph has been described and ready for execution, it is scheduled such that only data items with the same latency are fed into the nodes, and the nodes only execute if data are available and the nodes connected to its output are ready to accept the result. The scheduling can be done dynamically during runtime, with extra logic to guard each node, or in advance and statically during compile time.

2.2.2.1 Dynamic scheduling

The Manchester dataflow engine [7] bundles each data item with a label into a token, and a node is executed only if all of its inputs have tokens available and the labels match. The labels ensure that only data items are processed with the same latency, and allow to time share a node with unrelated data streams. The concept is flexible and exploits the currently achievable pipeline parallelism at runtime. Operations that require data not yet available block on an individual level until they can continue execution.

When implemented as general-purpose hardware, the bottleneck of this and similar designs is found in the distribution of work. For the Manchester dataflow engine, the supervision of node inputs and matching of the labels require content-addressable memory, which did not become available in sizes large enough for big applications. However, the architecture was successfully applied later to implement out-of-order execution in modern CPUs.

With custom hardware without resource sharing, the coordination between producer and consumer of data items can be implemented as a small FSM that governs the execution of each node. This requires at least a registered output that holds the data from the producer until every consumer has read it, as well as a signal to indicate availability of data set by the producer and an acknowledgement signal set by the consumers for the handshake protocol [12]. For small operations, the overhead of the FSM and extra signaling can be comparable in area to the actual logic. The individual node should therefore contain as much functionality as possible to keep the overhead small.

2.2.2.2 Static scheduling

Static scheduling aims to schedule the data dependencies of the dataflow graph a priori during compile time. This requires that the production and consumption of values are known before. Ideally, every node consumes and produces one data item with each clock cycle. No out-of-band signals are required in this case alongside the edges of the graph, because every cycle contains valid data and the whole system can operate in lockstep. The overhead is limited to observing the global inputs and outputs of the graph. The operation of the entire pipeline is then paused if a global input runs empty or an output stalls.

The architecture is very resource efficient, but puts major constraints on the logic it can describe. Most of the time, some nodes, such as reductions, will produce unneeded intermediate results, and therefore require extra logic to keep track when the end result is present at their output. For complex designs, the hardware parts of the design may approach dynamic scheduling again on top of static scheduling, leading to a convoluted dataflow description and a loss of abstraction.

2.2.2.3 Combined forms

In order to preserve the flexibility of dynamic scheduling and the resource efficiency of static scheduling, both approaches can be combined for dataflow computing on reconfigurable hardware. Nodes that can operate in lockstep are fused into a statically scheduled pipeline. These pipeline elements are then connected with buffer FIFOs at their inputs and outputs, and the fill level of the FIFOs is centrally supervised to decide whether a statically scheduled pipeline can run or must be paused. Depending on the abstraction level of the tool, the developer will either need to specify the boundaries of the statically scheduled subgraphs of operators, or the decision is made automatically and a set of options is provided for finer control.

2.2.3 Image Processing

Dataflow computing performs the same sequence of operations on every data item in parallel in a pipeline. An algorithm that processes a rasterized image and performs the same operation on every pixel is therefore well-suited to be implemented as a dataflow pipeline. At the input, a pixel is consumed by the pipeline with every clock cycle, and the result with the algorithm applied is presented at the output. The following primitives can then be combined to form a network of image processing operations.

2.2.3.1 Point operations

An image processing algorithm that operates only on individual pixels is known as a point operation. It is solely defined by a function that has the individual pixel as an input and returns a new pixel value. The function is then repeatedly applied on every input pixel until the entire image has been processed and a new image is obtained. For color images, the input and the output of the function is a tuple that describes the color value. Point functions define the most basic effects that can be applied to an image, such as negation, thresholding, conversions to grayscale, or other color adjustments. Point operations can often be found at the early stages of an algorithm and help normalize the image for the later stages.

Point operations are efficiently implemented with dataflow computing. The image can be streamed into a pipeline pixel by pixel and line by line, and the new image is received at its output. Point operators can be extended by an additional input in order to merge two images of the same size. If the second image is a delayed frame of the same recording, image arithmetic can be used to compare images with a time offset and use the difference to quantify changes over time.

2.2.3.2 Convolutions

Point operators are limited to individual pixels. Convolutions take into account the neighborhood of the pixel at the current position. They are image processing operators that move a small matrix, called kernel, over the image and multiply every kernel value with the underlying pixel values. The sum of the products then defines the pixel value of the output. The calculation is repeated for the entire input image until an output image of the same size is constructed. For an image that is defined by the lines and columns of matrix I, the convolution with (much smaller) kernel matrix S is defined in Equation (2.1) and yields the matrix I' of the resulting image [13]. The convolution kernel is sometimes also called stencil or mask.

$$I'_{x,y} = \sum_{i,j} I_{x-i,y-j} S_{i,j} \qquad (2.1)$$

The exact operation is defined by the values and the size of the kernel matrix. Equation (2.2) gives three example kernels [14]. S_{id} is the identity operation. The elements of S_{blur} approach a Gaussian distribution and the convolution will produce a blurred image. I_{sharp} is a Laplacian filter and is used to sharpen an image. Its effect vanishes for constant levels in the image, while intensity variations are emphasized in the result. Convolutions can be used for noise reduction, image enhancement, edge detection, and all other linear filters.

$$S_{id} = \begin{pmatrix} 0 & 0 & 0 \\ 0 & 1 & 0 \\ 0 & 0 & 0 \end{pmatrix} \quad S_{blur} = \begin{pmatrix} 1 & 2 & 1 \\ 2 & 4 & 2 \\ 1 & 2 & 1 \end{pmatrix} \quad S_{sharp} = \begin{pmatrix} 1 & 1 & 1 \\ 1 & -8 & 1 \\ 1 & 1 & 1 \end{pmatrix}$$

$$(2.2)$$

To receive an equally sized output image, the input image must be extended at its borders by half the width of the kernel matrix. This can be done by adding a zero margin, by extending the image periodically, or by mirroring it.

A convolution can be efficiently implemented for dataflow computing with a pipeline that keeps the image lines needed by the convolution kernel in a chain of fixed-sized FIFOs. For every clock cycle, one pixel is put line by line into the first FIFO. The pixel values needed for the convolution are tapped from between the FIFOs, multiplied with the corresponding kernel elements, and finally added to form the resulting pixel at its output. A CPU implementation that performs a convolution in place would need to save the pixel values in the neighborhood before overwriting them. Dataflow computing hides these performance-adverse register transfers in the FIFOs of the pipeline. The effect increases with the size of the convolution kernel.

Image convolutions can be extended to allow a more general neighborhood operation that is not limited to multiplications and a final addition. By introducing thresholds a neighborhood operation can also erode or dilate a shape in an image. Another use case is pattern recognition where the neighborhood of a pixel is compared with the pattern stored in the kernel matrix.

2.2.3.3 Reductions

At the end of an image processing pipeline, it is often required to not produce an output image, but quantify a certain feature of an image. In the section before, the center of mass of an object in the image was already presented as an example. The creation of a histogram reduces a picture to a vector of values that describe the frequency of an intensity level or color. In dataflow computing, these reductions can be calculated with a stream reduction if the access pattern on the input image can be made linear.

2.2.3.4 Operations with non-linear access patterns

Some image processing operations require a nonlinear read pattern for the input data or a nonlinear write pattern at the output. Geometric operations, for example, can only be implemented without buffering the entire image in RAM if the orientation of the image is preserved. The pipeline will then either discard values at its output to scale the image down, or insert cycles that do not read from the input for upscaling. Rotations, reflections, and general affine transformations require the image to be fully buffered in BRAM or a different randomly addressable storage, and the output image is started to be generated only after the input image was buffered. This disturbs the flow of data in the pipeline and increases the latency accordingly. Other common operations that require a nonlinear access pattern are bucket fill and the discrete Fourier transformations.

2.3 FPGA Hardware

Custom hardware is tailored to a specific problem and used where general-purpose hardware would not meet the required computing performance, such as sound processors or graphics cards. Other use cases are found in domains where general-purpose hardware cannot compete in terms of input bandwidth, latency, or cooling requirements. Traditionally, these computing solutions were implemented as fully customized hardware known as application-specific integrated circuits (ASICs).

Since the mid-1980s, FPGAs have been established as a third alternative that is located between general-purpose and application-specific integrated circuits (ICs) in terms of flexibility. FPGAs can be configured to change their functionality on the hardware level in a reversible process. The user can therefore choose from many different ASIC designs to be mapped on an FPGA model, as long as the resources of the FPGA are not depleted. For small series, FPGAs offer a way to avoid the expensive up-front cost required for the photolithographic masks of an ASIC production line and still provide the speed advantage of custom hardware. A second advantage over ASICs is given by the ability to update an FPGA with a new design to further improve the performance or to fix bugs later in the product life cycle, without changing the actual IC.

Since their invention, FPGAs have started to cover multiple use cases. Starting from ICs routinely used for glue logic, they have become a replacement for or emulator of digital ASICs and have gained traction as application accelerators in high-performance computing. To better understand FPGAs, we will first cover the internal setup and how it can be programmed and will then describe their usage as application accelerators.

2.3.1 Integrated Circuits

The first FPGA was introduced in 1985 by Xilinx Inc. [15]. The structure of a modern FPGA still follows the same principle: A large array of configurable logic blocks (CLB) is embedded into a wire network that routes the electrical signals (Figure 2.3). The binary operations of an FPGA are carried out inside of its CLBs, and the data consumed and produced by each of them is distributed

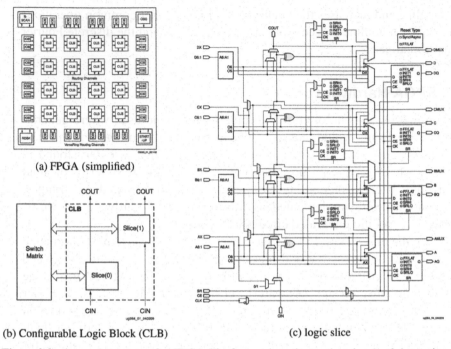

(a) FPGA (simplified)

(b) Configurable Logic Block (CLB)

(c) logic slice

Figure 2.3   Internal structure of an FPGA. The figures show the top three levels of abstraction of a Xilinx Virtex-6 FPGA as used in this book. Images: Xilinx [16, 17].

through the wire network. In this way, any logic function can be mapped by configuring the connections and the CLBs. The array of CLBs is surrounded by auxiliary units, such as input/output buffers (IOB), clock management (OSC), and other periphery needed for the configuration process and for communication with the outside world (Figure 2.3 (a)). For connections to a PCIe bus or a network interface, special high-speed transceivers are employed. Additionally, some FPGAs also feature entire CPU cores for heterogeneous computing within the FPGA.

2.3.1.1 Configurable logic blocks

The interconnects between the CLBs are configured in the switch matrix. Both CLBs and the switch matrix contain special static random access memory (SRAM) storage cells that are written during start-up and determine the function of the FPGA. This information is called the configuration of the FPGA. It is provided through its configuration interface from an external device or obtained from nonvolatile storage that sits next to the chip.

Each CLB is divided into two slices for Xilinx FPGAs (Figure 2.3(b)). They are connected to the switch matrices and to two adjacent CLBs for fast carry chains in adders and similar logic. An FPGA may contain several types of slices [16]. The internal structure of a logic slice is depicted in Figure 2.3(c) for a Xilinx Virtex-6 FPGA. It contains four lookup tables (LUTs) close to its inputs on the left, a set of multiplexers, and eight one-bit storage elements that may be used as flip-flops (FFs).

The SRAM content of each LUTs defines its truth table and is set during reconfiguration as well. A LUT can then be used to look up the result of any Boolean function with the same number of arguments at runtime. It substitutes the combinatorial logic between registers of an ASIC. The output of a LUT can either be directed to the carry chain by the multiplexers, to a FF, or out of the CLB to the switch matrix. The state of the multiplexers is also defined by the configuration, as well as the initial state of the FFs and their exact mode of operation.

All FFs have a clock-enable (CE) input. If it is set to low, the FF will not store a new value at the next clock cycles and keep the old value instead. This is a convenient feature when building pipelines that need to be paused repeatedly. All mutable state is contained in the FFs, and a low CE signal preserves the state until the pipeline will need to advance again.

Combined, LUTs and FFs provide combinatorial logic and storage within a well-defined unit. A modern FPGA contains up to hundreds of thousands of these slices. With just the design explained above, the logic of any digital ASIC could be mapped to an FPGA, given its resources do not get depleted. Because of the reconfiguration overhead, an FPGA needs much more transistors to build a logic primitive than an ASIC. With only LUTs and FFs, about 35 times the silicon area is required than an equivalent ASICs on average. In practice, however, the ratio can be brought down to under five times the silicon area with the support of special logic cores for common design units, for example multipliers and on-chip RAM [18].

2.3.1.2 Block RAM
Without dedicated FPGA resources for RAM, a design would require one FF per bit to be stored, and additional interface logic would be needed for read and write access through an address and data bus. To make better use of the silicon area, some slices contain LUTs that can be configured and combined to form distributed RAM. To store even more information with a low resource footprint, FPGAs embed special block RAM cores (BRAM). For the Virtex-6 series, an FPGA may have multiple thousands of BRAM cores with a size of 4.5 KB each [19].

Besides their combined ability to store megabytes of data directly on the FPGA, BRAMs can be accessed through two ports independently. That implies that in the same clock cycle, two addresses can be read or written. This is of particular use to build a first in, first out (FIFO) data buffer, where a write port inserts data at the front and a read port removes it at the end. This use case is common enough that some FPGAs support it directly with dedicated hardware support and offer signals to indicate when the FIFO is empty, full, almost empty, or almost full. A FIFO can then decouple different parts of an FPGA that are in a consumer–producer relationship and buffer unsteady data rates. The almost empty and almost full signals provide a convenient way to notify the producer or the consumer in advance that it will need to block soon. For pipelining, a FIFO provides the means to connect two pipelines that cannot operate in lockstep, and it provides the feedback to pause any of them when a stall occurs.

2.3.1.3 Digital signal processors

Adders and subtracters can be implemented with a low resource footprint on FPGAs with a dedicated carry lane. The hardware is then able to sum or subtract two numbers with fixed-point encoding per clock cycle. For a fully pipelined multiplication with N bits for each input, the resource usage grows with order $\mathcal{O}(N^2)$. Since multiplication is a very common use case, some FPGAs have special digital signal processing (DSP) elements available for area-efficient and fast multiplications. On the Virtex-6 series used here, a DSP slice consists of a 25×18 bit multiplier followed by an adder for fused multiply–add with two's complement integers. Each clock cycle computes the term $x = a \times b + c$ with correct rounding for the whole term. Common operations, such as matrix multiplications or convolutions, benefit from special DSP slices, and the silicon footprint of the FPGA configuration is reduced. Pre-adders are included to allow DSP slices to be cascaded for wider operators. Bit pattern recognition and feedback paths for accumulators further extend their use case to fast Fourier transforms, floating-point operations, and filter chains [19].

In the realm of image processing, DSPs are of particular interest for convolutions. A convolution on a rasterized computer image applies a filter kernel with a same-sized region in the image surrounding the current pixel, carried out repeatedly around every pixel of the image. It contains multiplications and additions and can make use of the fused multiply–add support. After the image has been processed with convolutions, an optional feature extraction stage will often require fixed-point or floating-point multiplications when the data are reduced to a few scalar properties, making further use of DSP resources. When the DSP resources are depleted, the remaining logic can still be used as a substitute, but may slow down the maximum achievable clock frequency, and hence reduce the throughput.

FPGAs may contain further hardwired cores to further increase the package density and customize them for their intended use. These cores can still be configured, but most of the logic implementation is fixed. A common core is a high-speed transceiver that connects an FPGA with a computer network or the PCIe bus of a host computer. Some product lines also feature entire CPUs for heterogeneous computing on the same silicon chip.

2.3.2 Low-Level Hardware Description Languages

The first FPGA, the Xilinx XC2064, consisted of an array of only 8×8 CLBs. At this size, the function of each CLB and the routing between the slices could still be performed on a sheet of paper. When ASICs and FPGAs grew more complex in the 1980s, it became apparent that hardware should be described with computer languages to increase productivity. At the same time, software was already written at a higher level and automatically translated to machine code. An equivalent tool flow for hardware was created for FPGA programming.

2.3.2.1 VHDL and Verilog

Two of the hardware description languages that were developed are still in wide use today: VHDL and Verilog. VHDL (VHSIC hardware description language) is based on the Ada language and was initially created as a means of documentation for digital ASIC designs. It became standardized in the year 1988 as an IEEE standard [20] and soon started to be used to simulate the behavior of the described ICs, too.

An example of a 16-bit register with two write ports described in VHDL can be seen in Listing 2.2. The code shown uses a subset of VHDL that can be transformed in a process known as logic synthesis into an actual configuration of an FPGA. The port map at the beginning declares which input and output signals the register uses. The process description below defines the functionality. The register may be written from one of the inputs at the rising edge of the clock signal (clk) by setting the write-enable (we) signal to high. The input is chosen by the select signal (sel) from either val_a or val_b. Otherwise, the value is kept. Entities like this one can instantiated multiple times and connected by code to form a more complex design.

```vhdl
library IEEE;
use IEEE.STD_LOGIC_1164.ALL;

entity reg16 is
  Port (clk    :  in STD_LOGIC;
    sel        :  in STD_LOGIC;
    we         :  in STD_LOGIC;
    reset      :  in STD_LOGIC;
    val_a      :  in STD_LOGIC_VECTOR(15 downto 0);
    val_b      :  in STD_LOGIC_VECTOR(15 downto 0);
    val_out    :  out STD_LOGIC_VECTOR(15 downto 0));
end reg16;
architecture Behavioral of reg16 is
signal val : std_logic_vector(15 downto 0);
begin
process (clk)
begin
  if rising_edge(clk) then
    if reset = '1' then
        val <= (others => '0');
    elsif (we = '1') then
        if (sel = '0')then
            val <= val_a;
        else
            val <= val_b;
        end if;
    else
        val <= val;
    end if;
  end if;
end process;

val_out < = val;
end Behavioral;
```

Listing 2.2 A 16-bit register in VHDL that can be written from two inputs. When the clock signal rises and depending on the input signals select (sel), write enable (we), and reset, the value in the register is either taken from one of the two inputs val_a or val_b, kept or reset [23].

Verilog serves the same purpose, but is based on the syntax of C. Both hardware description languages operate on the register-transfer level (RTL), meaning they are used to describe the flow of signals through combinatorial logic between registers that store the values. Both languages contain higher constructs such as loops with a variable number of iterations, but these can only be used for simulation and are not synthesizable to hardware. Because only a small subset is recognized (that also depends on the synthesis tool chain), a developer must know not only Verilog or VHDL as a language, but also the subtleties of its usage to eventually create working hardware. For efficient bitfile generation, they should also have a good knowledge about the internal structure of an FPGA. Otherwise, the configurable LUTs and FFs may not be used in the most efficient way. A description close to the hardware saves up to 50% of the resources when implementing a register alone, which is one of the most basic structures in a hardware design for digital logic [21, 22]. Hardware description languages that hide these details from the developer have been created since then, but VHDL and Verilog still remain the backbone of FPGA development.

2.3.2.2 FPGA design flow

After a design was written in a hardware description language on the register-transfer level, the logic must be translated into a configuration bitfile for the FPGA. This is done automatically by the design flow tools of the FPGA vendor. The process encompasses three steps, each building on top of the previous [22].

1. **Logic synthesis:** The abstract form of the design, typically written in VHDL or Verilog, is parsed, and special language patterns for state machines, multiplexers, or memory are recognized. Other optimizations check whether resources can be shared and will only need to be implemented once on the FPGA. The result is a description of the design as a set of logic elements and their connections. This is known as "net list" and describes binary and arithmetic operations for the combinatorial logic, registers for state, and logic cores that the developer specified directly in the source.
2. **Technology mapping:** The description from the previous step is transformed such that it matches the resources found on the target hardware. The combinatorial logic is mapped onto the LUTs and DSPs with their size taken into account, the registers are implemented as FFs, and the memory is built from the BRAM available or other resources. The result is specific for the targeted FPGA.

3. **Place and route:** Finally, the hardware primitives are placed on the FPGA and connected with routes through switch boxes. This steps will try to keep the wiring fast enough to reach the desired clock frequency by moving the logic on the chip until all critical paths are satisfied. The time needed for this step increases with the size of FPGA, the overall resource usage, and the desired clock frequency.

At the end of the tool chain, given all timing constraints could be met, a configuration bitfile will be produced that computes the source description on an FPGA for a certain voltage and temperature range. Contrary to software compilation, the step for placing and routing contains indeterministic searches for an optimum. For modern FPGAs with a high resource usage, the process can typically take from hours to multiple days.

2.3.3 FPGAs as Application Accelerators

We have seen that FPGAs require at least five times the silicon area for the same logic non-reconfigurable ICs. They also run at a lower clock frequency, typically with an order of magnitude in difference. Still, FPGAs have been shown to successfully accelerate programs that were previously executed on a CPU despite the initial penalty. The difference can be explained with the design constraints of general-purpose hardware. A CPU must execute general programs encoded in machine language, whereas the configuration of an FPGA can be customized to the wanted algorithm.

Figure 2.4 gives the floor plan of a core in the AMD Jaguar CPU, one of the few recent x86 CPUs where the area of its internal function units is available to the public domain. It depicts the size needed for auxiliary units that do not directly contribute to the computation and can mostly be omitted in custom hardware. About half the chip area of the CPU contains caches, buffers, and read-only memory, identified by their regular structure. Caches are usually smaller on FPGAs, because information about the memory access pattern is available in advance. The Ucode ROM is needed to translate the machine code into microinstructions to be executed by the other parts of the CPU. A machine language for general-purpose computing is not needed on an FPGA that computes one algorithm only, as it can be directly executed by the hardware.

The caches and buffers keep data and instructions close to the arithmetic and logic function units in a CPU. These are required to hide the latency of the working memory that compromises many clock cycles, during which the CPU

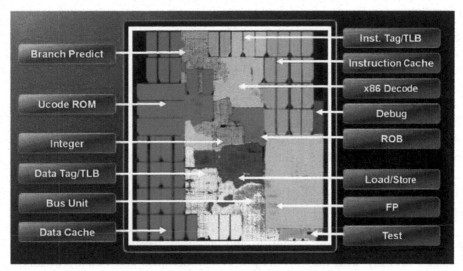

Figure 2.4 Floor plan of an AMD Jaguar core. Caches, branch prediction, and the reorder buffer (ROB) are dedicated units that are included to keep the pipeline running. Image: Advanced Micro Devices [26].

core would otherwise be waiting and idle. The data cache keeps parts of the RAM that are frequently accessed by the program, and the instruction cache makes the instructions of the program itself available for fast execution. The reorder buffer (ROB) modifies the order of parts of the program to avoid access conflict on the same storage location that would slow down the execution. Branch prediction finally applies heuristic to guess which path a program will take next to prepare the right commands for execution in the processor pipeline.

The core areas for integer calculations and floating-point (FP) logic then occupy only a small fraction of the floor plan. Most of the silicon area is dedicated to auxiliary units that accelerate data access for general-purpose computing. A significant part of the chip must also be powered off to prevent overheating. About 21% of a CPU at a 22 nm-silicon process must stay switched off at a time because of the much higher clock frequency and tighter integration [24, 25]. For custom computing on reconfigurable hardware, most auxiliary units can be omitted or at least greatly reduced, and in combination with the lower clock, FPGAs are able to beat CPUs in terms of speed, space, and electric energy for thoroughly ported applications.

2.3.3.1 Pipelining

The instructions of a program are executed at the center of an x86 CPU, surrounded by the caches and buffers. All but the most basic CPUs split execution of an instruction into sequential steps and perform every step in parallel to maximize throughput. The most basic pipeline for computers with a reduced instruction set (RISC) consists of five stages. First, the instruction is fetched from the instruction cache. It is then decoded into micro-operations in the second step and actually executed in the third one. If memory needs to be accessed, it is read or written in the forth step. Finally, the result of the instruction is written into a CPU register in the last step [27]. Modern CPUs have long pipelines with dozens of short stages to achieve a high clock frequency.

If the program has a linear structure and fits into the caches, the pipeline is filled all the time during execution and the clock frequency of the CPU correlates strongly with the throughput of instructions. The latency of an instruction in the pipeline is a multiple of the cycle duration between two consecutive instructions. Pipelining allows a computer system to hide this latency and keeps the throughput high. The instructions are not executed one at a time, but in parallel with an overlap as big as possible. For a perfectly running pipeline that produces one result per clock cycle, a pipeline increases the throughput by a factor equal to the number of pipeline stages.

For acceleration, it is important to estimate the different latencies of common operations in a computer system. Some latency numbers for common operations can be seen in Figure 2.5. They span nine orders of magnitude from a level-1 cache reference in a CPU to the (still barely noticeable) round-trip time of 100 ms on the IP network between the USA and Europe. The aim to mitigating latencies can be seen on every level of high-performance computer system: The CPU pipeline allows a frequency of instruction to be faster than the inverse of the latency of a single instruction. Branch prediction aims to keep the pipeline without empty bubbles. On a higher level, optimizations for longer latencies are built into the process scheduler of the operating system for concurrency. A process that waits for a locked mutex, disk storage or the network periphery will be de-scheduled and another process may continue execution on the same CPU.

Custom hardware does not have to support every random sequence of instructions. Instead, the program is known in advance, and the hardware is specially tailored. This allows the implementation of other types of parallelism and pipelining. Instead of pipelining the instruction flow, we can

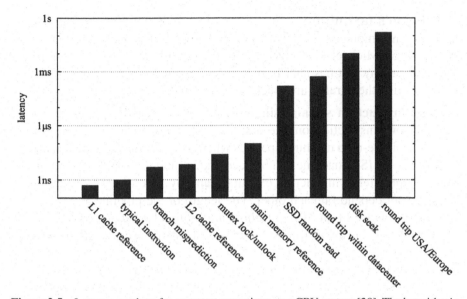

Figure 2.5 Latency numbers for common operations on a CPU system [28]. The logarithmic time axis spans nine orders of magnitude. With pipelining, the throughput of operations can be greater than the inverse latency.

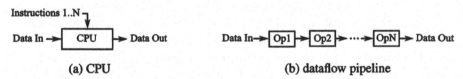

(a) CPU   (b) dataflow pipeline

Figure 2.6 CPU versus dataflow pipeline. In the CPU, the execution of the instruction flow is pipelined. If the stream of instructions is data-independent and repeatedly applied, the dataflow itself can be implemented in hardware as a pipeline on an FPGA.

pipeline the dataflow of a linear program without branching (Figure 2.6 (b)). The stages that fetch and decode instructions in a CPU can be removed, since the instructions are fixed. The execution stage in the CPU is reduced to only perform the current instruction, and the next instruction is appended physically on the FPGA to consume the result of the previous one. The execution of the program changes from an execution of instructions in time toward an execution in space, where each instruction in the program occupies its own silicon area. It is therefore also called "dataflow computing" or computing in space. Transforming programs with conditionals or complex loops requires extra effort and will be covered on its own in Section 3.3 "Pipelining Imperative Control Flows" (p. 44).

2.3.3.2 Flynn's taxonomy

Computers can compute in parallel, on either the instruction stream or the data stream. In 1970, Michael Flynn ordered the different kinds of parallelism in a concise taxonomy that has later been named after him [29]. It is still in use today to describe the parallel architecture of hardware.

- **Single instruction stream, single data stream (SISD):** A processor applies one instruction on one data item at a time. This is the mental model a single-core machine operates when it runs through an imperative program. A SISD machine can be accelerated hidden from the programmer by pipelining instruction execution and by reordering instructions that do not depend on each other. The length of the pipeline and hence the performance of an SISD machine are limited by data dependencies that occur when an instruction accesses data that were written by a previous one.

- **Single instruction stream, multiple data streams (SIMD):** Instead of applying an instruction to only one data item, a SIMD machine applies the instruction to a vector of data items. This capability can now also be found in previously SISD-only architectures. The most widespread example is the x86 processor architecture after the introduction of the streaming SIMD extensions (SSE) [30]. The second well-known member of the SIMD family is the graphics card with thousands of compute units that all perform the same instruction in parallel.

- **Multiple instruction streams, single data stream (MISD):** Here, multiple instructions are executed on a single data stream. A pipeline that processes a stream of data through multiple stages of operators can be seen as an MISD machine [31]. The number of (fixed) instructions processed per clock cycle rises linearly with the length of the pipeline.

- **Multiple instruction streams, multiple data streams (MIMD):** An MIMD machine can be obtained by parallelizing either a SIMD or a MISD machine a second time. For the first, MIMD is implemented by running a program in multiple SIMD cores. A pipeline (MISD) on the other hand can be parallelized by instantiating multiple instances, where each pipeline processes its own data stream.

The concepts of SISD, SIMD, and MIMD currently outnumber the MISD architecture by a big margin. This is likely due to the fact that the former three can be programmed with imperative, functional, or logic computer languages with few architecture-dependent additions. The SIMD architecture requires a transformation into a dataflow description to build the pipeline, and the mapping process from imperative code to it is still not readily available.

2.3.3.3 Limits of acceleration

Acceleration through frequency scaling has found its end in the last decade at frequencies of about 4 GHz. Since then, the focus has shifted toward parallelization. FPGAs that house thousands of operators running at the same time offer a way to further increase parallelism compared to the hardware in wider use today. The achievable gain through parallelism for a given problem was examined by Gene Amdahl, and the resulting law is known as Amdahl's law [32].

For a program that can be run in parallel except for a strictly serial part with fraction s, Amdahl's law describes the total acceleration a that can be achieved by increasing parallelism to a factor of N. The speedup will find its limit in the serial part; even an arbitrarily large factor of parallelism will not reduce its time. The total acceleration is calculated as in Equation (2.3). For a sequential part of $s = 5\%$, the maximum acceleration is limited to a factor of $1/s = 20$. The relation is shown in Figure 2.7.

$$a_{\text{Amdahl}} = \frac{1}{s + (1 - s)/N} \tag{2.3}$$

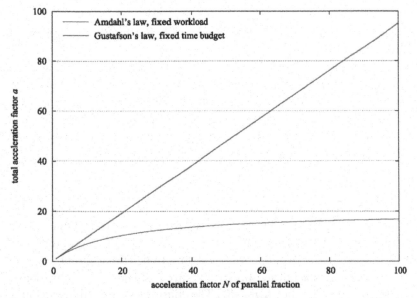

Figure 2.7 Amdahl's law and Gustafson's law. Given 95% of a program can be accelerated with factor N, the maximum total acceleration a depends on whether the problem size (Amdahl) or the total runtime (Gustafson) is fixed. In practice, a will be found between both predictions.

The resulting insight may seem very restrictive at a first glance. It implies a barrier for the maximum speedup for a given algorithm which cannot be overcome with parallelism. However, Amdahl's law acts on the assumption that the workload is fixed. In high-performance computing, this is usually not the case, and instead, the user will rather allocate a certain time span for computing. With a time budget of, e.g., one weekend or a fixed number of hours on a supercomputer, the maximum problem size is the primary interest of the user and researcher. Image processing is well-suited to adapt the complexity of the problem to the resources available, either by choosing a finer resolution, a larger area or volume, a longer time series, or a combination of the former.

John L. Gustafson noticed in 1988 that the time needed to compute the strictly serial part does not increase with the problem size in first approximation [33]. Hence, the workload that can be done in a fixed time grows proportionally with N for the parallel part $(1 - s)$, and the acceleration increases linearly with offset s (Equation 2.4). In the best case, the acceleration of the problem can be approximated with N. In practice, the maximal parallelization factor is limited by the communication overhead between the processing units, and the total speedup is found between the predictions of Amdahl's law and Gustafson's law.

$$a_{\text{Gustafson}} = s + (1 - s)N \qquad (2.4)$$

For a pipeline, the parallelism corresponds with the number of operators utilized at a time. An increased transistor count of an FPGA makes more CLBs available for configuration and therefore allows us to increase the length of the pipeline or the number of pipelines, leading to a corresponding speedup of the problem as a consequence. The serial part is usually caused by the time until data can be streamed into the FPGA from the periphery. The time spent between the start of streaming and the appearance of the first results is well below one millisecond and can be neglected for most use cases.

2.4 Languages

Since the creation of VHDL and Verilog as hardware description languages for FPGAs in the late 1980s, efforts have been taken to lift FPGA programming from the register-transfer level to a higher abstraction. The different levels of abstraction in hardware design are shown in Figure 2.8. The level of abstraction rises from the inner toward the outer rings. The chart was created first by Daniel Gajski and Robert Kuhn in 1983 [34].

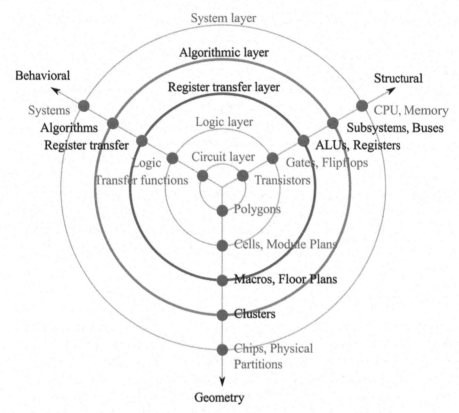

Figure 2.8 Gajski-Kuhn chart of hardware design. VHDL and Verilog operate on the complexity level of register transfers. High-level hardware languages move the description of the design into the algorithmic layer.

On the FPGA, the logic layer at the center of the chart is found at the actual wiring of the FPGA silicon chip and the implementation of its active elements. This layer is inaccessible to the programmer and its design is kept closed by the vendor. The next layer, the logic layer, contains the logic elements of an FPGA: LUTs, FFs, DSPs, and similar elements. VHDL and Verilog describe hardware on the layer above, the register-transfer level (RTL). Here, the behavior of a design is described as the transfer and transformation of logic values through combinational logic between registers. Despite the fact that both languages are able to describe algorithms with imperative statements on the higher levels, the subset that can actually be synthesized to hardware largely remains on the register-transfer level. The higher-level constructs, such as loops with a variable number of iterations, are limited to simulation. The development of large system is eased by intellectual property cores (IP cores) that act as reusable units of logic and range in complexity from simple adders to entire CPUs. These building blocks can then be combined by hand with glue logic on the register-transfer level.

In this section, an introduction is given to the efforts of describing hardware on the algorithmic level. The primitives found in these languages for constructing more complex designs already span multiple registers with combinational logic in between and can be assigned to the algorithmic level in the Gajski-Kuhn chart. The hardware designer typically combines mathematical primitives, such as arithmetic operators, and does not have to keep track of the data orchestration on the levels below. A higher-level description promises a shorter training period for the developer, gains in productivity, and, with FPGA silicon areas increasing, the option to design logic that would otherwise overwhelm the capabilities of a development team due to its complexity.

2.4.1 Imperative Languages

The 10 most popular computer languages can all be assigned to the paradigm of imperative programming [35]. Imperative programming languages describe computation as a sequence of statements embedded into control structures (e.g. conditionals and loops) that manipulate the state of a program. Among them are C, C++, and Java, and most programmers are able to write source code in at least one of them. VHDL and Verilog, on the contrary, do not follow the concept of linear execution of statements. Instead, all components are executed in parallel. VHDL and Verilog require special training, and a compiler that could automatically translate an imperative description toward a hardware description would make hardware programming available to a much larger group of (software) developers [36].

2.4.1.1 Handel-C

Handel-C is one of the earlier imperative languages for high-level hardware description for FPGA programming and inherits most of its syntax from C. It was created in 1996 at the Oxford University Computing Laboratory. The language features most of the data types from C except for unions and floating-point encodings, which are included as library functions. Integer types can be changed in size during variable declaration to make use of the flexibility of reconfigurable hardware. The language was also reduced by recursion, side effects in expressions and dynamic memory allocation [37].

```
par {
    f ();
    g ();
}
```

Listing 2.3 The par block in Handel-C. Statements within the body (f() and g()) are executed in parallel by the synthesized hardware. The functions are expanded at compile time and do not require a stack.

Plain C executes statements linearly. To describe hardware parallelism, Handel-C introduces the par block (Listing 2.3). At the beginning of the body, the execution flow splits and every statement (here f() and g()) is executed in parallel. The statements after the par block are executed sequentially again after every statement in the body has finished.

Information can be exchanged safely between parallel parts of a program through channels. These are FIFOs with a given capacity. When the FIFO is full, adding an element will block the sender until an element has been removed. The receiver will block accordingly when the channel is empty. Channels with zero capacity can be used to synchronize sender and receiver. The concept follows the notion of communicating sequential processes (CSP), which was developed at the same university and is known in high-performance computing as the underlying concept of the message passing interface (MPI).

Handel-C does not attempt to auto-parallelize code. The statements outside the par block are executed sequentially, with every assignment consuming one clock cycle. The delay statement can be used to introduce an idle cycle. With a combination of par and delay, pipelines can be created for MISD parallelism, although the programmer has to schedule the data streams manually.

2.4.1.2 Xilinx Vivado high-level synthesis

With the introduction of the Vivado design suite for its FPGAs of the 7 series in 2012, Xilinx also included an imperative high-level language to describe IP cores. Formerly known as AutoPilot, Vivado high-level synthesis (Vivado HLS) uses an extended subset of C for hardware design. The language omits, similar to Handel-C, the implementation of recursion, memory allocation, and system calls. Numeric data types can be refined to arbitrary precision, and a standard library for common mathematical operations is included [38].

The entry point of an IP core is given by a top-level C function that can be embedded into a C or C++ test bench or synthesized to hardware. The arguments of the function later form the inputs of the created IP core at the

RTL, and the outputs are derived from the function's return value. C arrays are translated to BRAM and add further inputs and outputs if they appear in the top-level function. Compilation produces VHDL, Verilog, or SystemC as an output that can then be embedded into a custom design of the same RTL language.

The compilation process creates a mixture of pipelines and controlling finite state machines (FSM) to map C programs to hardware. By default, the hardware executes the C statements in a strictly sequential manner, similar to Handel-C, but without functions' expansion. This can lead to hardware elements performing no operation for most states of the controlling FSM. Pipeline parallelism can be manually added, though, to address performance bottlenecks with pipeline directives that appear as C #pragma in the source. When a sequence of functions is pipelined, every function is translated to a hardware block and the blocks are connected with FIFOs, allowing each block to operate in parallel. When a loop is pipelined, the compiler will attempt to execute the body of the loop with pipeline parallelism if the data dependencies allow it. If pipelining fails, the compiler will generate a performance report to debug the data dependencies and resource conflicts. Other directives exist to unroll loops or inline functions.

Vivado HLS is one of the few tools that support both the dataflow and control flow domain. First results show that the resource footprint on the FPGA of the synthesized hardware resides remains low for image filter applications when compared to other tools for high-level synthesis [39].

2.4.1.3 ROCCC 2.0

The Riverside Optimizing Compiler for Configurable Computing 2.0 (ROCCC) is a C-to-hardware compiler that produces strictly pipelined hardware modules from the input [40]. It translates the most restricted subset of C-to-hardware modules in VHDL. By connecting the inputs and outputs of these modules via data streams, more complex systems can be created in a second step. For this, ROCCC strictly distinguishes between module code and system code. Both are written in C, but support different subsets. Module code operates on individual data items, while system code handles data streams through a C-array representation.

Owed to the requirement that all module code must be translated to a statically scheduled pipeline, the module code in ROCCC does not support short circuit evaluation. This kind of evaluation is a feature of C that causes the second operand of the binary operators && and || to be evaluated only if the value of the first operand requires it. These operators and, for the same reason, the ? : -operator have been removed from the language. Further restrictions in module code forbid generic pointers, non-for-loops, C-library functions calls, and nonpure functions. As with the systems described before, system calls, recursion, and dynamic memory allocation are neither part of the language [40].

Inside a module, the compiler unrolls all loops into a pipeline and must therefore know the iteration count at compile time. Outside, at the system level, the compiler will not unroll loops over the streams that connect the modules. Instead, the streams are represented as C arrays. The compiler determines which elements are accessed during an iteration and will determine the size of a stream window that covers all accesses. This stream window is then implemented with BRAM and caches the access to the data stream. The feature is especially valuable when an image convolution is to be implemented. The BRAM will automatically form a buffer for multiple lines such that the convolution kernel can access all covered pixel in parallel. The convolution kernel itself can be implemented as a module. In practice, care has to be taken to not let the stream window grow too far and consume all available BRAM resources [39].

ROCCC is open source, and its license allows the code to be modified. Researchers can study high-level transformations of the imperative source when mapping procedures, loops, memory access patterns, and other constructs to hardware without the need to build their own compiler [41].

2.4.2 Stream Languages

The semantic gap between imperative descriptions of an algorithm and its final form as a dataflow pipeline or FSM mapped into the silicon of an FPGA creates tensions between source code and hardware. From the compilers shown above, it can be concluded that, at the current state of the art, the semantic gap leads to either heavy hand optimization of performance-critical hardware modules or to a very restricted subset of C as the input language. In both cases, the developer must be aware that the target architecture imposes major restrictions to the design process.

Dataflow descriptions follow a different approach. The hardware developer is required to rewrite the algorithm into a pipelined form prior to coding. Afterward, the hardware is described as a directed graph with nodes that represent computations. The transformation is then more straightforward and limited to optimization, while the overall structure of the input description is preserved. As an advantage, resource usage and performance can be estimated from the input description, with every primitive corresponding to a certain resource footprint and maximum throughput. The actual process of porting the algorithm is moved to the beginning of the design process.

2.4.2.1 MaxCompiler

MaxCompiler from Maxeler Technologies uses a Java dialect to translate a dataflow graph to VHDL [42, 43]. Contrary to former efforts that aimed to translate Java to hardware [44], the programming language is only used to build the data structure of the dataflow graph. The Java program is not translated itself.

MaxCompiler was developed in Java as a successor of the C++-based "A Stream Compiler for Computing with FPGAs" (ASC) [ASC] and the domain-specific StReAm compiler [StReAm]. ASC introduced object-oriented design to describe hardware on the on the algorithmic level and the levels below in the Gajski-Kuhn chart. StReAm allowed the programmer to describe dataflow graphs in a natural way in C++ with operator overloading and templates.

```java
public   class AccuKernel extends Kernel {

    protected AccuKernel (KernelParameters params) {
      super (params);

      DFEType intType = dfeInt(32);

      DFEVar x = io.input (``x'', intType);
      DFEVar x2 = x * x;

      Params accuConfig =
              Reductions.accumulator.
              makeAccumulatorConfig(intType);
      DFEVar a = Reductions.accumulator.
              makeAccumulator(x2, accuConfig);

       io.output ("a", a, intType);
    }
}
```

Listing 2.4 Accumulator in MaxJ. The generated pipeline squares the values of the input stream "a" and produces the accumulation at its output.

The Java dialect used in MaxCompiler for the description of the dataflow graph is called MaxJ, a superset of Java that contains operator overloading to connect the graph edges with arithmetic operators. For simple, sequential-only algorithms that operate on a data stream, the result resembles imperative code. In Listing 2.4, the multiplication of x with itself acts on a stream of values, and the entire kernel object will get translated to a statically scheduled pipeline. More advanced structures, such as the shown accumulator, are instantiated as Java objects. Other node types for the graph include value nodes to produce a stream of constants, counters for bit pattern creation, and I/O nodes. Stream offset nodes shift a stream into the past or the future within a fixed number of clock cycles and are resolved into delays during scheduling.

Conditionals have to be implemented with the ? : operator or by explicit instantiation of a multiplexer. Loops must be scheduled mostly by hand and are created by connecting the output of a pipeline segment (identified by an object of type DFEVar) back to one of its input. The pipelines from the Kernel object are then embedded into a manager that connects the kernels with FIFOs to the host CPU, a network device on the accelerator card, or to other devices. A manager can also hold multiple kernels and will run them only if none of their inputs or outputs blocks. Multiple kernels with FIFOs in between can then be used for computing problems where the data cannot be piped in lockstep through a single pipeline,

The code is developed in the Eclipse IDE [45] with the support of the MaxJ plug-in (Figure 2.9). Running the MaxJ program will then produce the VHDL source from the dataflow graph. The VHDL will be passed to either the Xilinx or Altera tool chain afterward, depending on the vendor of the FPGA. The MaxCompiler library also produces an annotated version of the MaxJ source code that labels every line of code at the left margin with the number of FPGA resources it has occupied.

The Maxeler tools were used in this book for the implementation of the example applications. The included functions do not specifically support image processing, but the needed functions could be easily created. The usage of a general-purpose language for the generation of the dataflow graph makes the creation of libraries possible within the same language, where each MaxJ function generates a dataflow subgraph that can be parameterized and reused. Another use case occurs when the FPGA is not fully occupied after synthesis, and the graph can be extended for multi-piping by moving the responsible lines of MaxJ into a loop that instantiates a new subgraph with each iteration.

Figure 2.9   Screenshot of the MaxIDE 2011.3 integrated development environment. The dataflow graph is described with the MaxCompiler software library and MaxJ, an extension of the Java programming language. Synthesis can be started directly from the graphical user interface or through makefiles.

For testing, parts of the dataflow graph can be translated automatically to a C program that simulates the hardware with bit-level accuracy. Input creation and output verification are done in MaxJ as well. For program parts that do not fit the dataflow paradigm, the developer may use FSMs instead or revert to VHDL components.

2.4.2.2 Silicon Software VisualApplets

Contrary to the hardware description languages shown above, VisualApplets from Silicon Software does not use a text representation for hardware description, but a graphical user interface [46]. The user drags operators and links with the computer mouse to form a pipeline. The software then performs most of the synchronization, schedules the pipeline, and synthesizes the hardware with the Xilinx tool flow. Figure 2.10 shows the top-level view of a design.

Silicon Software offers special boards with Xilinx FPGAs that can be directly connected with a camera. The results of the image operation are

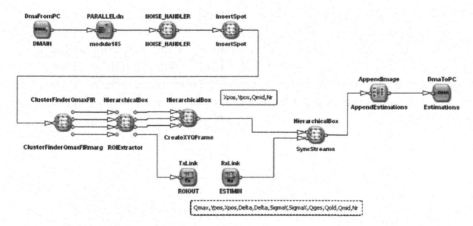

Figure 2.10 Graphical dataflow description in VisualApplets (Silicon Software). The image shows the top-level view of the pipeline implementation for localization microscopy. Each "Hierarchical Box" operator contains other operators. Image: Kirchgessner [48].

then shared with the host CPU through PCIe, similar to the accelerator cards presented before. It is also possible to feed imagery from the host CPU into the FPGA, as shown in the figure with the entry node "DmaFromPC."

VisualApplets distinguishes between two types of operators. O-type operators are simple operators for arithmetic and logic functions. They can be combined freely to form a point function and act on the individual pixel or perform basic reductions. M-type operators contain memory to buffer parts of a frame. They perform more complex tasks, such as convolutions and object detection. Due to the buffering, special care has to be taken by the user when the output of multiple M operators is to be used as input for the same operator. The differences in latency may otherwise cause a deadlock in the design [47].

The programming environment focuses on real-time image processing from a camera. The offered operators all rely on hidden out-of-band signals that indicate the end of line and end of frame when an image is transferred through the pipeline pixel by pixel. For data that do not constitute an image, the format must be changed such that it can be still processed. An array, for example, will need to be treated as an image with one line of pixels, and the pixel values encode the array elements.

One of the example applications presented in this book, the acceleration of localization microscopy, was carried out with VisualApplets by Manfred Kirchgessener within the scope of his diploma thesis [48]. He used the ability to feed data into the card from the host CPU in order to process the imagery that was collected before.

When this book was started, VisualApplets 1.4 did only support linear pipeline graphs. Loops could only be physically implemented by adding a daughter board on top of the accelerator card and routing the signals back into the FPGA. This has been solved in later versions.

3

Acceleration of Imperative Code
with Dataflow Computing

In this chapter, we will examine how the semantic gap between imperative code and a pipelined hardware architecture can be bridged. Imperative computer languages such as C and Fortran, but also object-oriented languages such as C++ and Java define a program with statements to control the flow of execution. CPUs directly support jumps in the control flow within their instruction set, and the development of computer languages has started with the go to statement. Imperative languages then moved on to conditionals and loops to organize jumps in the control flow in a more structured manner. Functions in functional languages and polymorphic member functions in object-oriented languages provide further abstractions to describe the control flow.

Dataflow computing, however, lacks these kinds of control structures. The aim of a dataflow description is to organize the flow of data instead of the flow of execution. The dataflow description of a program can be mapped well to an FPGA by transforming the dataflow graph to a system of pipelines in hardware, where each pipeline operates in lockstep. An implementation of control flow logic, however, would require state machines or simple processors that would serialize operations and could not fully utilize reconfigurable hardware, because at each time step, all resources not required by the current instruction would remain unused. Before laying out the paths for porting imperative code toward a dataflow description, we start with an overview of the techniques that arise naturally when processing streams in a set of pipelines.

3.1 Relation to List Processing

A stream of data in hardware is a serialization of values, where one value from a series is presented at a time in forward order. Without buffering, data that have been received in the past cannot be read again, and data from the future can only be accessed when we wait accordingly. The absence of free navigation in the data is a major constraint that immediately affects the design of a dataflow application. It contributes to the difficulties that arise when programmers from imperative languages learn dataflow computing for the first time.

In this section, we want to show that dataflow computing can be related to a concept that already exists in all popular computer languages. It will serve as an introduction to the semantics of dataflow computing. A stream of data in a pipeline has, beside from being navigable in only one direction, the property of immutability: Data can only be read from the output of the pipeline elements that form the inputs of the next pipeline element (Figure 3.1). The resulting data stream is a new stream and does not overwrite the input streams. The streams may be used as inputs by other pipeline elements as well and are not affected by their consumers. Dataflow computing on the pipeline level therefore copies immutable data at every pipeline stage.

The properties of unidirectional navigation and immutability can be found in software in *single-linked lists*. This data structure can be traced back to the first computer languages [49] such as C and Fortran. It is the primary data structure of Lisp [50] and can also be found as a first-class data structure in functional languages such as Haskell and declarative languages such as Prolog. Flat single-linked lists can be seen as a mapping of the temporal properties of a data stream into storage space. This picture enables us to go through the set of essential functions for list processing and state a dataflow implementation for each of them. Building on top of the pipeline primitives from Section 2.2, these solutions will then allow us to describe the translation of imperative code on a higher level in the following sections.

List processing is also closely related to image processing. Raw images and 3D voxel data can easily be streamed to and from an FPGA by storing them as a multi-dimensional array. These data are then serialized by accessing it row by row, column by column, and, for 3D volumes, layer by layer.

The relationship of dataflow computing and stream processing was already noted for general purpose dataflow machine with token labeling [51, 52].

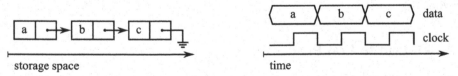

Figure 3.1   An immutable, flat, and single-linked list closely implements in space the semantics of a stream of data in time. Both formats can only be read in one direction.

Hardware pipelines that operate in lockstep need static scheduling that has to be carried out at compile time, rendering them less flexible. However, all concepts of list processing can be implemented, in some cases through the support of the memory controller. If the pipeline cannot be designed to consume or emit one data item per clock cycle, inputs and outputs have to be used that can be disabled for some cycles.

The standard library "Data.List" in Haskell [53] for processing of immutable, single-linked lists will be used as a guide to structure dataflow functionality in FPGA pipelines. C++ also provides a standard library for linearly traversable data structures in the <algorithm> library [54] that can be used similarly, but is less exhaustive and was written with a focus on mutable lists.

Of course, the data to be streamed to or from the FPGA do not have to be stored as a linked list. Instead, the data can be held in memory as an array to save storage space. The memory controller can then read the array linearly and pipe it into the FPGA, as well as write back the results accordingly. For the following implementations, we will assume that the input lists are stored in memory, either on the FPGA board or on the host computer, and the results are written back to memory. To implement nonlinear access patterns, the memory controller can optionally accept a stream of read or write addresses that define the storage location of each data item. An overview of such an FPGA system is shown in Figure 3.2.

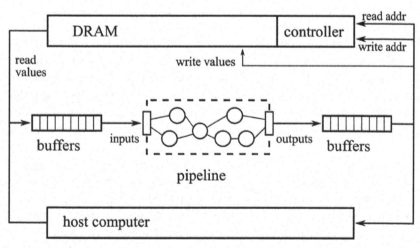

Figure 3.2 FPGA system overview with external memory and connections to the host computer.

When comparing Haskell's list processing functions with dataflow computing, we cannot pass around functions as first-class objects in dataflow computing. However, inlining can be used to statically modify the generic list processing facilities to a certain extent. List functions that are specializations of other functions already described have been omitted in the following comparison.

3.1.1 Basic Functions

head, last, tail, and init: These functions return the first element of the list (the "head" function in list processing languages), the last element ("last"), all but the first elements ("tail"), or all but the last elements ("init"). These functions may not seem useful on its own for dataflow computing, but are often necessary to exclude elements from the output of a pipeline. A pipeline generates one data item per clock cycle at its output, but complex pipeline system may produce intermediate results that should be filtered out before piping the output into the next pipeline or memory. An implementation on an FPGA consists of a counter and an output that produces values dependent on whether the counter counts the first element, the last element, or an element in between. By repeated application of "tail" and "init," a stream can be shortened by more than one element.

length: The length of a stream with a finite number of elements is usually stored in a register in advance from outside of the pipeline. It can then be compared to a counter for the "last" and "init" functions.

append: The append function takes two lists and appends the second list to the first one. For dataflow computing in a pipeline, we use two inputs, one for each data stream. At the beginning, only the first input is enabled and its stream is copied to the output. After the first stream has been processed, the first input is disabled, the second input is enabled, and the second input stream is copied to the output. A counter will be necessary to determine when to switch inputs.

Note that the combination of shortening a stream with multiple applications of "init" and the extension with "append" at its beginning by the same number of zero-initialized elements is equivalent to delaying a dataflow stream by introducing registers on the FPGA. This way, a stream element can be efficiently related to a previous one a fixed number of clock cycles ago. Compilers with automated pipeline scheduling also offer to advance one stream by a fixed number of clock cycles by actually increasing the latency of all other streams it affects.

3.1.2 Transformations

A list transformation creates a new list from an existing list. For a dataflow implementation, this means that the input and output of the pipeline correspond to the input and output lists, respectively.

map: A mapping creates a new list from the input list element by element. The mapping is defined by a unary function that takes a list element as an input and outputs the new list element. The return value does not depend on previous input values.

If the mapping function is already implemented as a pipeline, we can create a transformation in the dataflow picture by just piping the input values through the pipeline. The output then produces the transformed data. Input and output data can be stored in external memory and accessed linearly. With DRAM, this enables us to read from and write to memory with maximum throughput using burst mode.

reverse: The reverse function inverts the order of a linked list. In software, a reverse operation is implemented using the value of the first element as the last element of the new list and adding consecutive elements from the input to the head of the output list.

An implementation with a pipeline alone would need to buffer all data items before it could start to produce the last input data item as the first output data item. A pipeline cannot reverse a large stream of data due to the limited BRAM storage available on FPGAs. However, an FPGA implementation can utilize the storage controller for writing the data stream in reverse by sending a stream of declining storage addresses to the controller. To make use of the DRAM burst mode, the pipeline becomes a wire network that reads up to one burst per clock cycle, reverses it, and presents the result at its output where it will be written back.

intersperse: The intersperse function takes a list l and a scalar value. It inserts the scalar value behind every list element except the last one. On the FPGA, the stream input is disabled every other clock cycle, and instead of forwarding the input value, the scalar value is presented at the output. The scalar value can be fetched from a register. If needed, the implementation can be combined with the implementation of "init" to ensure that the output list is shortened by one item and the last element at the output is the last element of *l*.

3.1.2.1 Nested lists

The following functions from Haskell's Data.List library provide similar functionality to the functions above, but operate on nested lists. A nested list is a list that contains sublists as elements. If all sublists have a (small) fixed length, we can widen the pipeline and process one sublist per clock cycle. For large sublists, a hardware counter can be used to keep track of the position of each element relative to the containing sublist. A common example for this use case is matrices or images where every row is stored as a sublist.

For sublists with a variable length, a serialization protocol has to be used. For sparse matrices, for example, often only the nonzero elements are stored, reducing the number of values significantly. A protocol will therefore suppress all zero elements in a row and then compress the column and row coordinates of nonzero elements. An overview about storage formats for sparse matrices can be found in [55].

intercalate: The intercalate function works like the intersperse function, but inserts a list *k* instead of a scalar value in between the elements of list *l*. For small lists, *k* can be stored in the BRAM. The input of the pipeline can then be halted for the clock cycles where *k* is to be inserted. Often, the values of *k* do not matter and are inserted only to give the downstream part of the pipeline extra clock cycles. We can then generate *k* from zeros or repeat the last value read from *l*.

transpose: Matrices and images are transposed by swapping rows with columns. All columns must have the same length, as well as all rows. Therefore, we do not need a special format to nest rows into the data stream, but can rely on counters to keep track of the row and column number of each data value. Like the reverse function, we cannot transpose the whole matrix or image in the pipeline and need to rely on the memory controller to reorder the data. However, if we do not serialize the data by saving row after row, but store it as a series of square submatrices of the size of one memory burst, we can utilize burst mode similar to the reverse function. The pipeline should then be as wide as a burst and transpose it in a wire network.

subsequence: To output all subsequences of a given input data stream with fixed length *l*, we can send it repeatedly to the FPGA and use a counter with bit width *l* to generate a bit pattern for every repetition. The bits set to one in the counter value then provide a mask for which elements to output and which elements to suppress.

permutations: Like the previous function that reorders the data stream, the permutations function relies on the memory manager. From a data stream of length l, it generates all $l!$ permutations. Donald E. Knuth's Algorithm L [56] provides all permutations of a multi-set in linear time, enumerated by lexicographical order. It enables us to provide one data item per clock cycle in an FPGA implementation. For short input sequences, FPGA implementations exist that can provide an entire permutation at each clock cycle [57].

3.1.3 Reductions

A reduction consumes a (sub)stream and calculates a single scalar value from it. In functional programming, reductions that operate on lists are also known as folds. The name "fold" will only be used in this section as the term is also used in image processing for functions that produce a new image from a given one by applying a function to every pixel and its (limited) environment.

foldl, scanl: A left fold on lists applies a function to each element from left to right and uses the result of the previous application as a second parameter. A good example is an arithmetic accumulator with folding function $f(a, b) = a + b$, that calculates the sum of all elements when used with a zero starting value for the second parameter.

For streams, a left fold follows the natural order in which the stream elements are presented at its input. However, the applied function must be able to consume one element per clock cycle, which poses restrictions on the folding function. The dependence on the previous result creates an implicit loop with a latency that may not fit into one clock cycle. For summation, the accumulator will be covered in Section 3.3.3. Other examples for fold functions are binary "or" to accumulate all ones in a stream of bit masks, $\max(a, b)$ to obtain the maximum value of a stream, or a predicate in combination with "and" to check whether the stream contains at least one element that satisfies the predicate. The "scanl" function returns all intermediate results during accumulation and directly corresponds to the accumulator output on the FPGA. The "foldl" function returns the final accumulated value only. Hence, on an FPGA, the intermediate values have to be suppressed at the pipeline output similar to the implementation of the "last" function.

foldr, scanr: A right fold applies the fold function from the back to the front of a list. For dataflow computing, the stream therefore has to be reversed first. If the fold function is commutative and associative, a left fold can be used instead.

3.1.4 Generation

A list can be generated by repeating a constant or by running a loop that feeds the result of a function back into it as an argument. The concept is important in dataflow computing to generate auxiliary streams needed for computations on the input data.

repeat: The "repeat" function takes a scalar value as an argument and generates an infinite list by repeating it as a list element. This corresponds to the use of a stream of constants in dataflow computing, where the value can be hard-coded or stored in a register before the pipeline is started.

iterate, unfoldr: The "iterate" function takes a scalar value a and a unary function f to generate an infinite list. The first element is the scalar value, and the following elements are created by repeated application of f on the previous list element. If f is chosen to be the identity, function "iterate" will be equivalent to repeat. In an FPGA pipeline, the next trivial example is the counter without wrap and changeable initial value. It implements the "iterate" function with $f(x) = x + 1$. Similar to reductions, special care has to be taken due to the implicit loop that must produce one value per clock cycle, preventing the use of some functions. "unfoldr" implements the same functionality, but additionally lets f control when the list generation is to be ended. On the FPGA, it can be implemented by disabling the output of the pipeline.

cycle: This function generates list by repeating a given sublist infinitely. On an FPGA, the sublist can be stored in the BRAM if small enough and read repeatedly by connecting the output of a counter with wrap to the BRAM address port. For bigger sublist, the same principle can be used, but the sublist has to be stored externally in DRAM. Due to the linear address pattern and missing dependencies between the cycles, the function can be implemented by computing the linear address pattern a fixed number of clock cycles in advance, hiding the latency of memory access and making use of burst mode for DRAM.

3.1.5 Sublists

All sublist functions apply an explicit or implicit predicate function on each list element to determine which elements to return in one or more sublists and which elements to suppress. The order of the list elements is preserved. A hardware pipeline can implement this functionality by enabling the pipeline output for certain data items only, much like the implementations for "head," "last," "tail," and "init."

take, drop, splitAt: All functions take a list l and an integer n as arguments. The "take" function forwards the first n list elements and "drop" forwards all but the first n list elements. "splitAt" returns two lists: the first consisting of the first n elements of l and the second consisting of all other elements. For a hardware pipeline, all functions can be implemented efficiently by introducing a counter and comparing its output with the value of n to enable or disable the outputs.

takeWhile, dropWhile: "takeWhile" and "dropWhile" both accept a list and a predicate function. "takeWhile" returns a prefix of the input list until the predicate evaluates to false for the first time. "dropWhile" returns a suffix of the list, starting with the first element where the predicate evaluates true. Hence, both "takeWhile" and "dropWhile" produce disjunct lists when called with the same arguments.

An implementation for a hardware pipeline relies on a pipelined implementation of the predicate and an output that can be disabled. The one-bit output is then to be passed through a left fold operating with either $\min(a, b)$ for "takeWhile" or $\max(a, b)$ for "dropWhile." Finally, this data stream can then be used to control the output that forwards the input stream for all clock cycles when enabled.

span, break: Both functions take a list l and a predicate function as arguments and return two lists. The first list returned is the longest prefix of l where the predicate is true for "span" and false for "break." The remaining suffix of l is returned in the second list. The functions can be implemented in a pipeline similar to the "takeWhile" and "dropWhile" function, but with a second output that produces the otherwise discarded elements.

stripPrefix: "stripPrefix" takes two lists l and p as arguments. It checks whether p is a prefix of l, and, if true, returns l without the prefix. A pipeline implementation for streams of arbitrary length would need two passes over the data. In the first pass, p is checked whether it is a prefix of l. Depending on the result, l is stripped off the prefix in a second pipeline by enabling the stream output only after the prefix has been consumed.

For prefixes with a sufficiently small maximum length, the stream of p can be buffered, and only one pass over l is needed. If the suffix does not need to be forwarded immediately and l is stored as an array in memory, an alternative is to let the pipeline only check whether p is a prefix of l and shorten l in memory by updating its start address and length afterward.

group: This function transforms a list by grouping adjacent elements with the same value into tuples and returning them as items of a new list. In a pipeline, every element must have the same width due to a fixed number of wires. Therefore, a protocol is needed to serialize the tuples. As all elements in any tuple have the same value, we can encode a tuple by storing the value in the upper bits and size of the tuple in the lower bits. To calculate the size, we need to compare the present element at the input with the previous one and feed the signal into both a counter and the enable signal of the output. If the present and previous input elements are the same, the counter is increased by one and the output stays disabled. Otherwise, the counter is reset and its former value is sent to the output, alongside with the previous input element.

inits, tails: Both functions accept a list l as argument and return all possible prefixes ("inits") and suffixes ("tails") of l. On the FPGA, an efficient implementation of the same functionality can be achieved with the memory manager if l resides in DRAM, as the address pattern can be computed prior to memory access. The access pattern can be computed with two counters, where one counter creates the addresses and the second counter increases the start value ("inits") or the wrap value ("tails") of the first counter every time it wraps by one. The stream from memory then contains the desired prefixes or suffixes.

isPrefixOf, isSuffixOf: The "isPrefixOf" and "isSuffixOf" functions both accept two lists l and s and return whether s is a prefix or suffix of l, respectively. In hardware, "isPrefixOf" can be implemented by comparing the elements of l and s in a pipeline stage and accumulating the result with an "and" reduction to check whether all pairs of elements are equal. A counter is used to emit the state of the accumulator when the stream of s has been consumed. For "isSuffixOf," a similar implementation can be used that skips the first $(\mathrm{length}(l) - \mathrm{length}(s))$ elements of l before it compares the remaining pairs. If s is sufficiently short, its elements can be held in either DRAM or BRAM storage.

isInfixOf: This function has the same signature as "isPrefixOf" and "isSuffixOf" and checks whether s is an infix of l. For short infix lists, s can be stored in length(s) registers, where the nth register is compared to the stream delayed by n clock cycles. A pipeline would then first read and store s in its registers before processing l. The registered values are compared with their corresponding stream values. A stream reduction stores whether all all comparisons evaluated true for at least one clock cycle. The state of the reduction forms the return value at the end.

It is also possible to search for matches of regular expressions instead of infixes if the regular expression is known at compile time. A regular expression can be translated to a nondeterministic finite automation, which can be reshaped into a deterministic FSM [58] and implemented on an FPGA. The FSM consumes one value per clock cycle. It can therefore be embedded into a pipeline and scheduled without extra logic. The translation of regular expression into FSMs finds its limit in the number of states required. It increases exponentially with the length of the (very convoluted) regular expression in the worst case.

3.1.6 Searching

elem, notElem, lookup: All functions take a list l and a value v as input. "elem" returns true iff v is an element in l, while "notElem" returns true iff v is not an element in l. "lookup" compares v to every first element in a list of pairs and returns the second element of the pair of the first match. "lookup" can therefore be used to implement an associative array with a list of pairs where the first element acts as the key.

For a pipeline implementation of "elem" and "notElem," the present value of the input stream representing l is compared to v and the Boolean result is fed into a "and" or "or" reduction. The last element of the reduction is then the wanted return value. The "lookup" function is similar to the "elem" function, but with some extensions. The pairs of the input streams are encoded as the lower and upper bits. The comparator only compares the lower bits with v, and the upper bits are fed into a register that stores the value only when the return value of the reduction switches from false to true.

find: "find" takes a list and a predicate and returns the first list element the predicate accepts. For a hardware implementation, a pipelined version of the predicate is needed. The list is forwarded to the pipeline output, which is only enabled when the predicate returns true for the first time. The control logic can be implemented by combining the output of the comparator with a stream accumulator based on the "max" function, which in turn is connected to an edge detection to enable the output only once at most. If "find" must also return a result for non-matches, a second output may be added that submits the value of the accumulator in the last clock cycle.

filter, partition: "filter" takes a list and a predicate as arguments and returns all elements of the list where the application of the predicate evaluates true. In hardware, the list input is connected to the pipelined version of the predicate and also forwarded to the output. The predicate filters the stream by controlling the enable signal of the output.

"partition" shares the same signature as "filter," but additionally returns a list of elements that cause the predicate to return false. The implementation can built on top of the solution for "filter" by adding an additional output that is also connected to the input stream and only enabled when the first output is disabled, creating two disjunct streams.

3.1.6.1 Indexing lists

All searching functions can be easily modified to return the position of the searched value instead of the value itself. To do so, a counter is needed that tracks the position of the present value in the data stream. When the value is found in the stream, the counter value is presented at the output.

3.1.7 Zipping and Unzipping

zip: The function "zip" takes two lists as arguments and returns a list of pairs, where each pair is composed of the elements of the input lists in forward order. A pair of values can be encoded in hardware by abstracting the wires of both inputs as a single entity. No logic is needed to implement the function in a pipeline.

unZip: This function is the reverse of "zip" and splits a list of pairs into two separate lists. Similar to "zip," no logic is needed to assign the wires to separate streams in a pipeline.

zipWith: This function accepts two lists much like "zip" and a binary function. The output is the resulting list of the element-wise application of the input lists. In a hardwire pipeline, a basic binary operator has the same semantics and can be combined with other operators to implement more complex operations. Any pipelined binary operation is therefore a hardware implementation of "zipWith," and variants that accept more than two input lists can be implemented in a pipeline accordingly.

3.1.8 Set Operations

Set operations treat lists as sets from set theory. A pipeline can only access a limited window in the data stream and hence requires an additional constraint that the elements of the input streams are to be ordered for set unifications, intersections, and differences. The next section will explain how to sort lists in hardware if needed. The semantics of the comparator used for sorting must be the same for the set operator.

nub: The function returns a list where all duplicates have been removed from the input list. The same semantics can be implemented in a pipeline by forwarding the input stream to the output only if the present and the previous stream values are different. The input list has to be sorted beforehand.

delete: This function accepts a list l and a value v and deletes the first occurrence of v from l. In hardware, the present stream value is compared to v. If both values are equal for the first time, the output is disabled; otherwise, the stream is forwarded. The same control logic as for "find" can be used to ensure a deletion at only the first occurrence.

difference: The "difference" function accepts two lists l and m as arguments and returns a list that contains all elements of l that are not present in m. In hardware, both input streams have to be sorted. If both inputs present the same value, no value is forwarded to the output. If the value at the input of l is less than the present value of m, the value of l is consumed and forwarded to the output. Otherwise, the present value of m is consumed and no output is produced.

This function cannot be implemented in MaxCompiler as a pipeline due to the fact that the value of the read-enable signal for the inputs depends on the input value and MaxCompiler requires the enable signal to be known three clock cycles in advance. However, an FSM can be used instead that controls the inputs by comparing their values and checking for empty or stalled FIFOs in a single clock cycle. This implicit cyclic dependency between input controls and input values puts stress on the timing of the design because the comparator logic cannot be pipelined. On modern FPGAs, a comparator can be run at 150 MHz for data streams with up to 64 bits and at 100 MHz for up to 128 bits.

union: This function returns the unification of both lists *l* and *m* provided as arguments, returning a new list that contains all values that are elements of *l, m,* or both. In hardware, both input streams have to be sorted, similar to "difference" and "intersect," and can be implemented as an FSM. If both inputs present the same value, both inputs read in the next value and forward the present value to the output. Otherwise, only the smaller value is consumed and forwarded to the output.

intersect: The function returns the intersection of two lists: Only elements that are present in both lists are part of the resulting list. In hardware, both input streams must be sorted, and the function has to be implemented as an FSM. The present value of both lists must be forwarded to the output only if the present values are equal. If not, the input with the smaller element must be advanced and discarded.

3.1.9 Ordered Lists

sort: This function takes a list as an argument, sorts it following a stable algorithm, and returns the ordered list. The function cannot be implemented with a single pass through a pipeline for streams of arbitrary length, since the pipeline has only access to a limited window of data items at a time. All comparison-based sorting algorithms have a time complexity of at least $O(n \log(n))$ for *n* data items [59]. Hence, a pipeline would need to process the data in at least $O(log(n))$ passes.

A hardware implementation can make use of merge sort, which is stable and requires only few logic for merging. Similar to the implementation of the set operations, an FSM can be used that forwards the smaller value of its inputs to the output. A naive implementation would repeatedly read in the data from memory in blocks with a length of a power of two, merge them, and write them back to memory with the support of a custom address generator. After $\lceil \log_2(n) \rceil$ iterations, the data would then be sorted. A more advanced hardware implementation on FPGAs speeds up the initial passes with a sorting network and then processes multiple merges in parallel [60].

insert: "insert" accepts a sorted list and a value. It inserts the value into the list and preserves the order. The function is a special case of "union" with a one-element list as second argument and can be implemented accordingly.

3.1.10 Summary

Functions that consume and produce one data item per clock cycle can be combined freely in a longer pipeline to form more complex functions. These are the functions "map," "scanl," "zip," "unzip," and "zipWidth." Functions that generate data streams can be used at the beginning of a branch of the pipeline and replace pipeline inputs otherwise fed from the host or DRAM: "repeat," "iterate," "unfoldr," and "cycle" are members of this class. Functions that remove elements from the data stream must be wired to the enable signal of the output and can be serially combined by feeding the individual enable signals into an "and" node. Examples are the "head," "last," "tail," and "init" functions as well as the sublist functions and all predicate functions. All other functions namely the set operations "difference," "union," and "intersect," the sort function, and "permutations" can be combined using separate pipelines that are connected through buffer FIFOs with each other, the DRAM memory, or the host computer.

For image processing, the "map" function directly implements point functions on rasterized image data. Point functions modify each pixel independent of its neighborhood and are needed for color transformations. The second most important operation on image data is a fold that creates a new value for each pixel by applying a function f to its former value and the value of its environment. Depending on the function and the size of the environment, a fold can implement FIR filters and other convolutions on image data as well as erosion and dilation for morphological image operations [61]. On the FPGA, the data stream of pixels is delayed by multiple clock cycles to create streams that represent the pixels in the environment (see "append" and "init"). A pipelined version of the function f is then applied (see "zipWith") and calculates the new pixel values, one per clock cycle.

3.2 Identification of Throughput Boundaries

To successfully accelerate an algorithm with hardware support, it must be understood in detail. Setup and cleanup code must be identified and separated from the computational core of the algorithm; the bounds of size and frequency of the data streams between its distinct parts must be estimated or measured; and the inherent bottlenecks that limit its performance on CPUs must be considered as a starting point for acceleration.

Profiling the program gives quantitative data about the behavior of the software. It increases the knowledge about the runtime behavior and provides insight which functions are part of the computational kernel. The different load levels of the system's components indicate whether the speed is held back by the CPU or the memory and I/O system. The following paragraph lists the different performance bottlenecks, ordered by increasing latency. The latency of the individual components is shown in Figure 2.5.

compute bound: The program is limited by the speed of the CPU and would benefit from an increase in processing capacity. Programs that are compute bound have a low ratio of stalled CPU cycles during execution. The processor either accesses the memory and I/O subsystem seldom, the working set of data fits well into the fastest cache, or both. If the underlying algorithm can be formulated as a pipeline, a CPU-bound program can be accelerated by exploiting the massive parallelism of the FPGA hardware.

cache bound: A program that is bound by the cache would benefit from a larger cache. Depending on the access frequency, the program can be limited by the first level cache or one of the larger, but slower caches behind. On the FPGA, the performance of a cache-bound program can be increased by holding the working sets in BRAMs and registers. The data can then be accessed in parallel at the frequency of the computing components, given that the algorithm has been parallelized before in one or more pipelines.

memory bound: The bandwidth of applications that are held back by the speed of memory access is throttled by up to three orders of magnitude for random access when compared to cache-bound allocations. Random memory access that causes cache misses in the entire cache hierarchy must be eliminated as far as possible for a speedup. For both FPGAs and CPUs, the data layout should be changed to reduce the size of the working set, by either tiling the data and computing on the individual tiles separately in the cache, or switching from random access toward mostly linear access, taking advantage of the burst mode of DRAM.

   Further, data can be compressed if it contains redundant information to reduce the size of the working set. Lossy or nonlossy compression virtually increases the memory bandwidth at the expense of the computing resources dedicated to compression and decompression.

I/O bound: The execution speed is bound by I/O if the program gets stalled when waiting for network or disk I/O. These operations can reduce the bandwidth by up to three orders of magnitude, too, when compared to a memory-bound program. Few options remain for acceleration with computing hardware alone: Compression can be used similar as for memory-bound programs, and algorithms can potentially be changed to compensate a reduction in data bandwidth with better processing.

3.2.1 Profiling in Software

Acceleration is measured in terms of improvements to computing time. To identify the compute kernels of a program for acceleration, it is therefore necessary to measure the time that is spent in every function. Software profilers rely on the compiler, which inserts additional instructions during compilation that record when each function is entered and left at runtime. Profiling information can also be sampled statistically by inspecting the instruction counter regularly at the tick of a high-frequency timer interrupt. From the value of the instruction pointer, the currently executed function can be traced back. Modern profilers use both methods to generate an overview of program execution [62]. For languages such as Java that are compiled to byte code, the virtual machine provides the recording infrastructure [63].

To identify the most time-consuming functions, the profiler subtracts the runtime of all callees from the runtime of the caller before ranking them. Otherwise, the main function would appear at the top with an execution time of close to 100%, while the actual computation is carried out by the much more interesting subfunctions. The functions that appear on the top then need to be examined individually to research the limiting factor of their execution time and their potential acceleration.

A section of an example output of the gprof tool [62] is shown in Figure 3.1. The program that was profiled is the C version of the analysis for localization microscopy, presented in Chapter 4. All functions listed were ported to the FPGA and process image data on the pixel level, except for the constructor for character strings. The cumulative time of a function also includes the time of the functions it calls, while the self time excludes it. Functions may benefit from acceleration if they run a long time, such as the subtr_and_update_bg() function that is applied to an entire image and removes its background, or because they are short, but called very often, such as the min() and max() functions that compute their mathematical counterpart. Finally, the constructor for character strings is called only once and belongs to the setup of the program, making it unsuitable for dataflow acceleration.

Software profilers can also record memory allocations, system calls, and other runtime data. While this information is crucial to acceleration

% time	cumulative seconds	self seconds	calls	self ms/call	total ms/call	name
58.57	0.89	0.89	1998	0.45	0.64	tiff_image16_ref::subtr_and_update_bg (tiff_imag
14.48	1.11	0.22	40919040	0.00	0.00	short const& std::min <short> (short const&, shor
10.53	1.27	0.16	40919040	0.00	0.00	int const& std::max <int> (int const&, int consts
7.90	1.39	0.12	1993	0.06	0.09	clfinder::find(tiff_image16_ref&)
3.29	1.44	0.05	1	50.01	50.01	global constructors keyed to TypeString (unsigne
2.63	1.48	0.04	26101	0.00	0.00	estimator::estimate(tiff_image16_ref&, roi, dou

Listing 3.1 Output of the software profiler gprof. Manipulations of entire images belong to the longest operations, while the utility functions min(a, b) and max(a, b) are called most often. Setup code is called only once.

in software, memory allocators and system calls to the operating system stay in the realm of the CPU and are not ported to a hardware pipeline. For hardware acceleration, a different set of data is more important to decide which code could benefit the most: bottlenecks in computing can be widened with MISD or SIMD parallelism, and bottlenecks in data access caused by cache and memory latencies are removed by pipelined data access with preknown access patterns. Both kinds of bottleneck are largely invisible to applications in user space. Support from the operating system and the CPU hardware is needed to obtain these performance measurements with high accuracy.

3.2.2 Profiling the CPU System

Modern CPUs contain hardware counters in the performance monitoring unit (PMU) that can be programmed to track performance events. Every time the event occurs, which can be a cache miss, a memory reference, or any other performance critical trigger, the counter is incremented. When a specified maximum count is reached, the counter is reset and a high-priority interrupt is emitted to notify the operating system [64]. During process profiling, these counters can be read through a kernel interface by user space profilers and combined with the process context. By setting the maximum count to a low value, the performance event of interest can be monitored with high resolution and the locations in the source code that triggered the event are known through the value of the program counter. A high maximum count, on the other hand, reduces the overhead and avoids distorting the measurement. Tools such as "perf" [65] for Linux adjust the maximum count during runtime such that a constant user-defined sample frequency is reached.

The capabilities of the PMU depend on the processor it is part of. However, all popular architectures support performance events to track the performance of the memory hierarchy and arithmetic computations. This information is crucial for the identification of bottlenecks and to classify the boundness of the application. With this information, an FPGA engineer can then already estimate the speedup of hardware acceleration.

In Listing 3.2, the result of a run of "perf list" with default parameters is shown. The program profiled is the already optimized C version of the analysis for localization microscopy, which was profiled in software in the previous subsection. With the additional information, we can now determine that the hardware bottleneck by going through the list of possible bounds from bottom to top. An I/O bound program would show a large number of page faults, as would a memory-bound program. The data, however, indicate that the number

```
Performance counter stats for './main data/07.tif':

87161.268021 task-clock-msecs      #     0.992  CPUs
       12042 context-switches      #     0.000  M/sec
         388 CPU-migrations        #     0.000  M/sec
       20499 page-faults           #     0.000  M/sec
234550320502 Cycles                #  2690.993  M/sec
359160744267 instructions          #     1.531  IPC
   406552059 cache-references      #     4.664  M/sec
     6487996 cache-misses          #     0.074  M/sec
87.826693930 seconds time elapsed
```

Listing 3.2 Output of the CPU profiler perf for an already optimized program. Few cache misses and more than one instruction per clock cycle indicate that a program is compute bound.

of page faults is only 20,499 during an execution time of 87.8 s. The next bound is the cache, which is missed in less than 2% of all cache references. Since last-level cache has 7% of the latency of memory (Figure 2.5), the program must be compute bound. This is also indicated by the number of instruction per clock cycle (IPC), which is larger than one, and proved by the acceleration that was achieved with an FPGA system later. Further optimization with vector instruction, of course, could easily increase the number of IPC and make the program cache bound.

3.2.3 Profiling Dataflow Designs

After the compute kernels have been identified and while the program is being ported to the FPGA, it is often necessary to also profile the execution in the reconfigurable hardware. When a statically scheduled pipeline produces one result per clock cycle, its throughput can be calculated easily by multiplying the data width times the clock frequency, and its latency is given by the number of pipeline stages. However, pipelines often show a more complicated behavior. The pipeline can run empty if the input data are not provided fast enough by the host, the memory controller, or another pipeline. It may also be forced to wait when one of its outputs stalls. Last, but not least, controlled inputs and outputs can be used to introduce cycles where the pipeline does not consume or emit values.

These potential performance bottlenecks can be detected by introducing counters that monitor when the pipeline was running empty, stalled, and computing. The pipeline designer is then able to read out the according registers after a run and relate the values to the total number of clock cycles. If one of the pipelines is causing the drop in performance, it can be examined for multi-piping, and I/O bottlenecks can sometimes be widened by compression or a modified encoding.

Figure 3.3 shows the performance counters obtained through the "maxdebug" tool for a pipeline built with MaxCompiler. The performance counters can also be embedded into a stream between kernels to monitor the fill level of FIFO buffers. The ratio of the amount of words transferred versus the total number of clock cycles informs about the utilization of the output and input that heads and tails the buffer. The number of stalled clock cycles indicates a limitation at the receiver, and the number of empty cycles relates to a slow sender. Since FIFO buffers form the interfaces of statically scheduled pipelines, their performance numbers are valuable indicators to track down a performance bottleneck in hardware designs. In the final design, these counters can be omitted to return their resources to the application.

3.3 Pipelining Imperative Control Flows

After the performance bottlenecks have been identified, we can start to translate the compute kernels of into a pipelined hardware design. In this section, we will assume that the software was written in imperative languages such as C, Fortran, or C++ which are popular choices for high-performance applications. It can be shown, however, that all computer programs that run on a CPU can be ported to an FPGA: A CPU only consists of combinatorial logic and registers, and both elements are available in pipelined computing as well. Combinatorial logic can be build with stream operators, and registers are covered by stream reductions. Given that the FPGA provides enough resources, the CPU can then be instantiated as a soft core and the program can be run on it.

The aim of this section is to not only port software to hardware, but also port it with a focus on efficiency. A soft core processor cannot provide an efficient implementation due to an increased consumption of silicon area and a decreased clock frequency by a factor of about an order of magnitude when compared to a legacy CPU. We have seen in Section 2.3 that general-purpose hardware is elaborately designed to perform well with any kind of code. The complex control and cache logic that surrounds the arithmetic core of the processor becomes dispensable for FPGA hardware that is configured

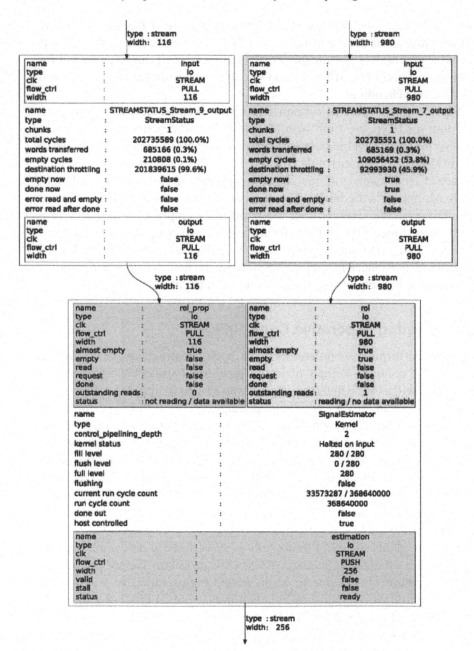

Figure 3.3   Profiling of a pipeline with MaxDebug. The values of the stream counters are shown after a run for a statically scheduled pipeline (kernel SignalEstimator) and for two of its inputs.

to execute one program only. The now spared silicon area can then be used to perform the actual calculation.

In the best case, the program can be translated to a single pipeline that runs in lockstep and contains few control logic. Here, almost all hardware resources can be utilized immediately to perform the actual calculations of the program. To accomplish a design that follows this principle as close as possible, we need to examine how the semantical concept of imperative computing can be translated toward such a pipeline. The building blocks of imperative computing are sequences of statement expressions, conditional and loop statements.

Imperative programming contains further concepts [66]. Procedures and function calls maintain a structure. Both can be inlined if they do not perform recursive calls. Recursion can be mapped to sequential code with a loop and an explicit auxiliary stack. Object-oriented code, which is part of the imperative code family in a wider sense, relies on polymorphism. When needed for a calculation, the type identification is to be made explicit in a first step and function dispatch can then be performed with a conditional statement based on the type information.

More advanced concepts in imperative programming are rarely found in compute-intense kernels of applications. These are exceptions, which can be substituted with an additional return value for each function that encodes an arising exception. The concept is especially known from system programming in C where functions return an invalid value on error and set the global variable "errno," and more recently from the Go programming language [67]. Continuations, which can be implemented in C with setjmp() and longjmp() to mimic tail call optimization, are even less common. Last, but not least, the "goto" statement is "considered as harmful" [68] to good software design and can be substituted with conditional and loop statements.

The structured program theorem [69] proofs that indeed, any computable function can be rewritten to only contain sequences, conditionals and loops. Each control structure and its translation to a dataflow description will be examined below. For declarative languages, which form a superset of logical languages and functional languages a straightforward, low-level translation to the dataflow paradigm can be very difficult: Both of them make mandatory use of lazy evaluation and garbage collection, two concepts that require a uniform memory layout in CPU implementations and are contrary to stream processing. On a higher level, the probably most popular members of both language families are Prolog and Haskell, which both make heavy use of processing of single-linked and immutable lists. When rewritten accordingly, the processing of lists is then close to dataflow computing again, as we have seen in Section 3.1.

3.3.1 Sequences

The simplest building block of an imperative computer language is the statement expression. These are arithmetic and logical expressions that produce a value and optionally assign it to a variable. Contrary to variables in mathematics or functional languages, the values of variables in imperative languages can change after initial definition. Expressions may also have side effects. These cause other variables than the variables on the left-hand side to change its value, too. In C and related languages, side effects are caused by the increment (i++, ++i) and decrement operators that alter the value of a variable before or after it has been read. Statement expressions can call functions that may have side effects as well on global variables or variables passed by reference.

We have seen that dataflow programming corresponds to the processing of immutable list. Mutable variables which are a core concept in imperative languages are not supported directly in the dataflow paradigm. Instead of altering a variable, its value is read or copied, fed into an operator and the resulting value is assigned in the initialization of a new dataflow variable. The semantic gap can be bridged by transforming sequences of assignments to static single assignment (SSA) form, where each variable is assigned a value at exactly one place in the source code. Variables V that are assigned multiple times are substituted with a set of new variables V_i (for $i = 1,2,3, \ldots$) until all variables are assigned only once [70].

In Listing 3.3, an example sequence of statement expressions in C is shown inside of a function that calculates the length of a 2-dimensional vector. The function length() computes the length of a 2D vector. In its body, the variable s is assigned a value twice, first at declaration and a second time in the line after. By introducing a new variable s2 for the second assignment and by renaming s to s2 in the following source code, the double assignment was removed in Listing 3.4.

```
float length(float x, float y) {
  float s = x * x;
  s += y * y;
  return sqrt(s);
}
```

Listing 3.3 length() with multiple assignments to s.

```
float length(float x, float y) {
  float s = x * x;
  float s2 = s + y*y;
  return sqrt(s2);
}
```

Listing 3.4 length() following static single assignment form.

After the code has been transformed to SSA, the dataflow graph of the algorithm can be drawn immediately. The nodes in the graph represent the operators in the source code. The edges indicate the transfer of values and are directed from variable declaration toward variable readout. SSA ensures the acyclic property of the graph as every variable is assigned only once and cannot be read before it has been declared. For expressions with more than one operator, the expression is mapped to the dataflow graph according to its syntax tree and defined by operator precedence. For the length() function, the dataflow graph is shown in Figure 3.4. It is also the graph of the resulting pipeline.

The generated pipeline is able to process one vector per clock cycle and keeps the utilization of each processing element at 100% as long as both inputs can provide the data stream and the output does not stall. Between the operators and inside of some of them, registers are inserted to keep the longest path in the combinatorial logic short and the clock frequency high. These registers increase the latency, but do not slow down the computation for fixed clock frequencies. Complex operators like the square root may have a latency of more than one cycle, but can still process one data item per clock cycle. Sequences of assignment expressions can therefore be implemented with very high efficiency on an FPGA. More complex pipelines than the one shown here will be discussed in Chapter 4.

For the pipeline to run embedded in a hardware design, few extra logic has to be instantiated that stops the pipeline when one of the inputs runs empty while enabled or one of the outputs stall. In this case, the pipeline can be stopped by halting each register that connects combinatorial logic. On the

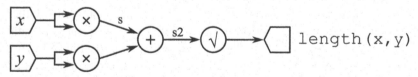

Figure 3.4 Dataflow graph and pipeline of the vector-length function.

FPGA, the registers are built from flip-flops that usually already come with a clock-enable signal input. Setting all of them to the low logic level will therefore halt the entire pipeline.

The dataflow path reveals that a software implementation could make use of instruction-level parallelism: The multiplications that square x and y in the example do not depend on each other and can be calculated in parallel. A compiler could analyze the dataflow as we did and use a vector (SIMD) instructions, especially if the vector had a greater number of components. The hardware implements an extra level of parallelism as it also computes operations where the input of one operation depends on the result of a previous one (MISD). The implication leads to a different understanding of state in sequential code and pipelined dataflow computing. A pipeline computes l instances of the same computation in parallel, where l is the latency of the pipeline. Therefore, the parallelism grows with the number of pipeline stages, and C-like code can be parallelized by creating deep pipelines that represent the algorithm first (MISD) and afterward multiplying the pipeline (MIMD) until most of the resources of the FPGA are occupied. Translated back into a software paradigm, the parallel computation of a sequence corresponds to list processing, where the list elements visualize the timely flow of data streams in storage space.

The transformation presented here from sequences of assignments to dataflow pipelines is straightforward and can be implemented in a compiler or software library to be performed automatically. In the next section, we will examine conditionals. The translation of branching builds on top of the translation of sequences.

3.3.2 Conditionals

Conditional statements in control flow languages like C divert the flow of control depending on the result of a previously calculated condition. On the processor level, a conditional is a selective jump of the program counter. In this section, we will examine the conditional jump that does not lead to a loop in the flow of execution. In C and related languages, it is implemented as the if-else statement or the switch statement. The switch statement acts a syntactic sugar and can be expressed with multiple if-else statements. Other occurrences of conditional execution in C are the ternary operator "? : " and the conditional "and" (&&) and "or" (||) operators that evaluate the following expressions dependent on the result of the first one. These operators can be rewritten with if-else statements and by introducing a new variable for the value of the subexpressions.

Listing 3.5 shows the abs () function that returns the absolute value of a signed integer. If the input value of x is negative, it is inverted; otherwise, x is returned. The hardware implementation of the same function can be seen in Figure 3.5. Contrary to the software implementation that will execute only one branch depending on the condition, both branches must be implemented in hardware and consume FPGA resources, and the correct result is selected afterward by a multiplexer. The branching itself is done by wiring, and it is only the reunion of the data streams that requires LUTs to implement the multiplexer.

Note that the conditional creates two meshes bordered by three paths in the dataflow graph, and all three inputs of the multiplexer must have the same latency to ensure the data arrives synchronized. In the example, a delay element was inserted before input "a". These delays, implemented using extra registers or FIFOs in BRAM, need to be added until each path has the latency of the longest one. As a side note, the predicate ($\bullet < 0$) can be implemented as a single wire from the most-significant bit for two's complement encoding of x and will then require a delay as well. For more complex meshes in the dataflow graph, the minimal number of delay registers can be calculated automatically from its directed acyclic graph (DAG) by optimizing a system of linear equations [71].

Using multiplexers has the disadvantage that all branches need to be implemented in hardware, but only one result is used later, while the other

```
int abs(int x) {
  if (x < 0) {
    return -x;
  } else {
    return x;
  }
}
```

Listing 3.5 Description in C of the abs () function.

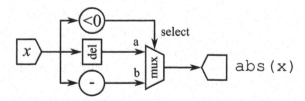

Figure 3.5 Dataflow graph and pipeline of the abs () function.

intermediate result is discarded. For conditionals with short branches like the shown abs() function, this is tolerable. Multiple options exist when the overhead becomes too large.

- **Separate synchronous pipelines per branch:** In particular, if the conditional has one branch only, a second pipeline for this branch connected with a buffer can help save resources. Only if the condition evaluates true, the data are passed to the second pipeline. The speed of the second pipeline can then be adjusted according to profiling data from the condition. The second pipeline can either be clocked slower, yielding a higher freedom in circuit design, or make use of resource sharing and consume multiple clock cycles per data item. An example of the latter can be seen in Section 4.1, where the second kernel requires 49 clock cycles to process one ROI. Loosely coupled pipelines move the scheduling of the design from static to a more dynamic one at the expensive of extra buffers and control logic.

- **Finite-state machines:** Pieces of code that contain a large number of conditionals, such as compiled regular expressions, parsers, and lexers with long switch statements, can be implemented as an FSM instead of a pipeline and share resources between states. The mentioned use cases consume one data item per state transition and can therefore seamlessly integrated into a pipelined design. FSMs are also well-suited for conditional-heavy control logic that would become convoluted in the pipeline paradigm.

- **Partitioning between host CPU and FPGA:** Code branches that are executed seldom or only at the start or the end of a calculation and are responsible for only a minor part in the total computation time can be left on the host CPU to save FPGA resources. These are typically setup and cleanup tasks, and code branches that are connected with low data bandwidth with the rest of the system.

- **Multiple bitfiles:** Code branches that are executed repeatedly in a loop, but where the condition evaluates to the same value for long sequences that last hundreds of millisecond, can benefit from separated FPGA configuration. When the condition changes, the configuration is changed, too, and no resources are wasted for unused branches. Configuration, however, takes tens to hundreds of milliseconds and is only an option if the condition changes seldom. Partial dynamic reconfiguration can decrease reconfiguration times.

- **Multiple FPGAs:** If the system consists of multiple FPGAs, different bitfiles can be used to specialize them for the multiple, but disjunct paths of the dataflow. To fully utilize all chip resources, the option requires static scheduling and a constant ratio of the individual branch decisions in short timescales and may increase latency due to additional buffering.

- **Changes in the algorithms:** An FPGA design can greatly benefit from algorithmic co-design as shown in the application chapter (Chapter 4). Good sources for algorithmic alternatives with few branches can be found in the software world: Compiler architecture avoids conditionals, too, as they slow down the processor pipeline. The abs() function, for example, can be calculated without a branch with the functions "xor," "and," and subtractions alone [72]. A second resource is time-independent cryptography functions that avoid conditionals as well.

The options presented here for conditionals still have to be chosen by hand. Only the static scheduling with multiplexers as described first can be automated by transforming the code to SSA. In SSA, the Φ function that assigns values to the variables at the end of the conditional, dependent on the branch taken, is then substituted with a multiplexer [70]. Up to date, the best options can only be selected after profiling as described in Section 3.2 and understanding the algorithm.

3.3.3 Loops

With sequences and conditionals alone, the state in the registers between pipeline elements is flushed out when the pipeline advances. Hardware that needs to implement functions with an arbitrary long impulse response must make use of feedback loops to keep state in the pipeline. The output of an accumulator, for example, depends on all previous input values since reset and therefore needs a feedback loop to add the current input value to the output value from the previous clock cycle.

We will first focus on loops that can be unrolled or where the loop body can be calculated in a single clock cycle. These loops are straightforward to implement and fit well into synchronous pipelines that process one data item per clock cycle. In the first case, the loop vanishes and the dataflow graph processes all iterations in parallel. In the second case, a loop with a latency of a single clock cycle for the loop body is constructed in hardware by simply connecting the output to one of its inputs. Loops with larger loop bodies cannot always be converted and will form loops with a latency of more than one clock cycle. These loops result in a more complex scheduling that is known as loop tiling and will be examined last. An overview of all presented loop techniques is shown in Figure 3.6.

Note that all long-running software contains loops in one way or the other in the control flow. Otherwise, a gigahertz CPU would traverse the instructions of a program with the size of multiple gigabytes within a few seconds and terminate. Therefore, the computational kernel of a long-running program must contain at least one loop.

3.3.3.1 Loop unrolling

Loop unrolling can be used if a loop has a fixed number of iterations. Given that the number is small enough to not exhaust too many chip resources, the loop can be unrolled to a sequence of statements, and the resulting acyclic pipeline can be laid out on the FPGA. The resource usage is determined by the number of resources consumed by the loop body times the number of iterations. Besides resource constraints, loop unrolling does not impose any restrictions to the source code to be applied; moreover, information can be passed between non-neighboring iteration more easily than in uprolled form. Loop unrolling removes the loop and therefore puts no constraints on the latency of the pipeline of the loop body. It comes with the advantage of maximum throughput and a simple integration with the other parts of the design, but may cause high resource usage for nontrivial loops.

An example can be seen in Listing 3.6. The function is known as bit population count. It counts the number of ones in the binary representation of

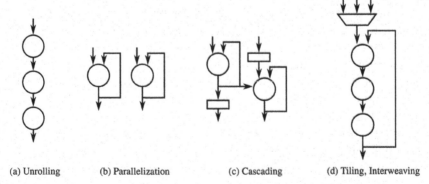

(a) Unrolling   (b) Parallelization   (c) Cascading   (d) Tiling, Interweaving

Figure 3.6   Loop implementation techniques for static synchronous dataflows. Circles indicate operations with a clock latency of a single cycle. These operations can form a loop by simply connecting the output to one of its inputs. Otherwise, the loop must be unrolled, dismantled into loops with single-cycle latency, or manually scheduled.

```
int pop_cnt(uint16_t a)
{
  int cnt = 0;
  for(int i = 0; i < 8 * (int) sizeof (a); i++) {
    int16_t mask = 1 << i;
    cnt += (a & mask) ? 1 : 0;
  }
  return cnt;
}
```

Listing 3.6 Bit population count in a loop in C++.

an unsigned integer by applying a moving mask to the integer. As the number of bits in an unsigned integer is fixed, the number of iterations is known in advance and the loop can be unrolled. Data structures that are generated from the control variable i become constants, and as a consequence, the masks in the example become a set of constants as well.

The corresponding pipelined hardware version of the unrolled algorithm is shown in Figure 3.7(a). It features 16 adders to calculate the bit population count, assuming standard integer encoding. Because addition is an associative operation, the adders can be re-arraigned into a tree shape to save flip-flops in the delay registers (Figure 3.7(b)). The lower stages of the adder tree have been substituted with three times three lookup tables, limiting the number of needed adders to only two.

3.3.3.2 Loop parallelization

Applications that process large amounts of data from memory following a certain access pattern or from I/O channels in a linear order usually apply the same processing steps to the data iteratively. In imperative software, the process is described using an outer loop or, for multi-dimensional data, a group of nested loops. In hardware, however, the outer loop is not immediately visible, as the I/O subsystem takes care of feeding data into and fetching results from the pipeline.

A loop in software affects its hardware counterpart wherever values are written in one iteration and read at a later one, which represents a special case of a read-after-write dependency. In this case, a feedback loop must be implemented that holds the updated value and routes it back to the pipeline stage where it has to be read. Loops in software will generally pass the updated value to the very next iteration. In a hardware pipeline, the affected stage will process the next iteration in the next clock cycle. Therefore, the latency of the

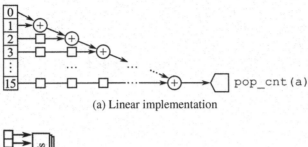

(a) Linear implementation

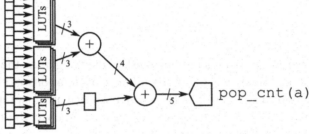

(b) Tree-shaped implementation

Figure 3.7   Pipelined version of the bit population count function in hardware. The tree-shaped implementation saves delay registers and adders.

hardware feedback loop is limited to a single clock cycle and must have both the input and output connected to the same pipeline stage. Hence, if a loop in software does not write to a variable (or any other storage location) after the same variable has been read in the loop body, the generated hardware will not contain a feedback loop. In this section, we will focus on loops that lead to feedback paths, since the former kind can be handled like a sequence of statement expressions.

Loop parallelization and loop cascading are two design patterns that help to rewrite feedback loops such that they are compatible with the one-cycle latency constraint. If both techniques cannot be applied due to technical reasons or excessive resource usage, loop tiling can provide a solution that supports higher latencies.

Loop parallelization aims to untangle multiple independent read-after-write dependencies. Contrary to automatic loop parallelization with SIMD in software compilers [73], these loops can still be mapped to hardware. An example is shown in Listing 3.7 that computes the arithmetic mean and the variance from a data series in a naive way. While the first for-loop merely accumulates the data for the calculation of the mean, the body of the second loop is more complicated: Two subtractions, one multiplication, and finally an addition on top of var certainly will not fit into a single clock cycle without

```cpp
pair<float, float>
mean_var(int x[], int N) {
  float mean = 0;
  float var = 0;

  for(int i = 0; i < N; i++) {
   mean += (float) x[i];
  }
  mean = mean / N;

  for(int i = 0; i < N; i++) {
   var += (x[i] - mean) * (x[i] - mean);
  }
  var = var / N;


  return std::make_pair(mean, var);
}
```

Listing 3.7 Calculation of the arithmetic mean and variance in C++.

severely slowing down the clock frequency of the entire pipeline. Moreover, the second for-loop depends on mean, the result from the first loop.

To calculate both loops in parallel, notice that the variance of a data series can be calculated without incorporating the mean at every step. Instead, we accumulate the values of x_i^2 in linear time and subtract the value of $\bar{x}^2$ at the end (Equation 3.4). The transformation removes the data dependency between the loops.

We can then accumulate x_i and x_i^2 in parallel to compute $\bar{x}$ and Var(X) (Listing 3.8). The final division by N and the subtraction of the squared mean $\bar{x}^2$ at the end can either be processed on the FPGA or on the host computer, given the data sets are large enough to not slow down the overall throughput. The hardware implementation then looks like Figure 3.6(b) with integer adders as operators and with an input stream of x_i for the first accumulator and an input stream of x_i^2 for the second accumulator. A counter could be used to only enable the output for the last result from the accumulators.

```cpp
pair<float, float>
mean_var(int x[], int N) {
   int x_acc = 0;
```

```
int x2_acc = 0;

for(int i = 0; i < N; i++) {
  x_acc += x[i];
}

for(int i = 0; i < N; i++) {
  x2_acc += x[i] * x[i];
}

float mean = (float) x_acc / N;
float var = (float) x2_acc / N - mean * mean;
return std::make_pair(mean, var);
}
```

Listing 3.8 After rewriting, both for-loops could be calculated independently.

$$\bar{x} = \frac{1}{N} \sum_{i}^{N} x_i \qquad \text{(arithmetic mean)} \tag{3.1}$$

$$\text{Var}(X) = \frac{1}{N} \sum_{i}^{N} (x_i - \bar{x})^2 \qquad \text{(variance)} \tag{3.2}$$

$$= \frac{1}{N} \sum_{i}^{N} x_i^2 - \frac{2\bar{x}}{N} \sum_{i}^{N} x_i + \bar{x}^2 \tag{3.3}$$

$$= \frac{1}{N} \sum_{i}^{N} x_i^2 - \bar{x}^2 \tag{3.4}$$

The example was taken from the application for localization microscopy that is presented in Section 4.1 and common for center-of-mass feature extraction. The conversion of the formula for the variance was rather simple, but cannot be done automatically with today's tools. The translation of imperative software to hardware remains a process where the underlying algorithm of the software must be codesigned with the hardware.

3.3.3.3 Loop cascading
In some loops, a read-after-write dependency may be present that cannot be removed by rewriting the algorithm. If the loops can still be simplified into

loops with a body that is short enough to be calculated in a single clock cycle, loop cascading can be applied. Here, one loop acts as a producer of a stream of values that is consumed by one or more secondary loops. Loop cascading can therefore implement algorithms in hardware that could not be ported with loop parallelization alone.

In Listing 3.9, a software implementation of a 128-bit accumulator for unsigned integers is shown. Mainstream processors cannot sum 128-bit integers natively. The function performs the summation by splitting the integer into two 64-bit unsigned integers. After the lower halves of both integers have been added, a possible overflow is detected and the higher halves are summed alongside with the carry flag of the first operation.

On the FPGA, we certainly could instantiate a single 128-bit adder, but the most significant bit of the result will depend on all less significant bits in the bit vector of the inputs and will lead to a carry logic with long paths. To not reduce the clock frequency of the chip too much, it is advisable to similarly split the hardware accumulator into two (or more) adders for high and low bits.

The read-after-write conflict in the function `acc_uint128` prevents us from splitting the loop into two independent parts in software. The carry bit

```cpp
struct uint128 {
  uint64_t high, low;
};

uint128 acc_uint128(uint128 a[], int N)
{
  uint128 sum, prev = {0, 0};

  for(int i = 0; i < N; i++) {
    sum.low = prev.low + a[i].low;
    uint8_t carry = prev.low > sum.low ? 1 : 0;
    sum.high = prev.high + a[i].high + carry;
    prev = sum;
  }

  return sum;
}
```

Listing 3.9 An accumulator for unsigned 128-bit integers in C++.

is written by the first addition of the lower bits and read in the second for the higher bits. In hardware, however, the loop can be segmented into two accumulators, with a wire for the carry bit connecting both (Figure 3.8). Delay registers allow the second loop to operate on the data one clock cycle later than the first loop, such that the carry bit is available in time. The enable signal at the output allows only the last result to be forwarded. It has been omitted from the figure for simplicity and can be implemented with a counter and a comparator, where the counter consists of an accumulator with its input set to constant one.

The example for-loop unrolling (Listing 3.6) could also have been implemented with loop cascading. The first loop contains a left shift and provides the bit mask, while a second loop counts all occurrences where the "and" operation of the mask and the input does not result in zero.

Both loop parallelization and loop cascading require logic that can provide its result with a latency of a single clock cycle. These are additions and

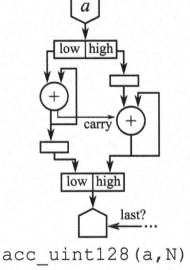

`acc_uint128(a,N)`

Figure 3.8 Accumulator implementation with loop cascading.

subtractions of (sufficiently short) integers and fixed-point numbers, and all binary operations that operate locally, such as the binary primitives not, and, or, shift, rotate, or related functions. Logical functions that are more complicated enforce a throughput-limiting reduction of the clock frequency owed to long critical paths.

3.3.3.4 Loop tiling

A loop in software cannot be transformed to a hardware loop without extra measures if the body of the loop must be transformed to a pipeline with $l > 1$ stages. A feedback loop in hardware would calculate l iterations of the loop body in parallel, and data could not be exchanged between these iterations, splitting the loop into l pieces. If loop unrolling cannot be used due to resource constraints and both loop cascading and loop separation are prevented by data dependencies, loop tiling can act as an option to generate resource-efficient translation into hardware.

As an example, a floating-point accumulator that calculates the sum $\sum_i x_i$ in hardware cannot be implemented with a floating-point addition that would have a latency of only a single clock cycle. To add a floating-point number, several steps must be carried out in a linear manner. The summands must at least be normalized, the exponents must be aligned, the mantissas are to be shifted accordingly, and only then, the mantissas can be added and the floating-point number is encoded according to the standard again. All these steps require one or multiple clock cycles in latency to keep the clock frequency at an acceptable level. On a Xilinx Virtex-6, an IEEE floating-point addition is synthesized to a pipeline that has a latency of $l = 12$ clock cycles for single and double precision. Feeding the result of the addition back as an input to create an accumulator would actually create an accumulator that splits the input data into l independent and interleaved substreams and will accumulate them independently of each other. An illustration of the issue can be seen in Figure 3.9.

To solve the problem, an auxiliary adder pipeline could be appended that sums the l interleaved substreams up again into a single result stream. The resource usage of such an adder would be linear with latency l, and therefore, solutions have been developed to keep l small [74]. These solutions can only work for operations like addition that are commutative, as the order of execution is modified. Number-crunching applications, however, often apply

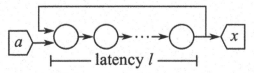

Figure 3.9 An accumulator implemented naively with a series of operations for floating-point addition and a feedback loop would calculate the sums of l interleaved substreams, with $x_j = a_{j-l} + a_{j-2l} + \ldots + a_{j\bmod l}$ instead of the desired result $x_j = a_j + a_{j-1} + \ldots + a_0$.

a single instruction stream to multiple independent streams of data (SIMD) in nested loops and can therefore benefit from a technique called "loop tiling." The term was first coined by Maxeler Technologies for static scheduling in hardware [75] and is similar to loop tiling in software compilers. In software, a loop gets reordered and split into multiple subloops by the compiler to improve cache reuse and enable vectorization [76]. For hardware design, the loop is rescheduled into work units of l independent calculations.

Equation 3.5 gives an expression to calculate the total gravitational force F_i of a mass m_i at position $\vec{x}_i$ as part of the n-body problem by summing up all contributions from all remaining masses m_j. We are interested in all force vectors $\vec{F}_i$ to calculate the change in momentum of all masses later in an exemplary simulation with discrete time steps. The problem arises in all simulations with long-range interactions like galaxy simulations or molecular-dynamic simulations.

$$\vec{F}_i = \sum_{j \neq i}^{n} Gm_i m_i \underbrace{\frac{\vec{x}_i - \vec{x}_j}{|\vec{x}_i - \vec{x}_j|^3}}_{=:\vec{f}(\vec{x}_i, m_i, \vec{x}_j, m_j)} \quad i = 1 \ldots n \qquad (3.5)$$

The time complexity of calculating all $\vec{F}_i$ is of $\mathcal{O}(N^2)$ as for every calculation of F_i, the properties of all N bodies have to be read from memory. In software, the problem can be solved using two nested loops (Listing 3.10). These nested loops provide the opportunity to re-arrange the control flow.

The example reads the input stream composed of masses m_i and positions $\vec{x}_i$ twice with index i and j, applies a reduction on it with the overloaded C++ operator $+=$, and returns an output stream F_i. Listing 3.11 implements the algorithm with equal semantics, but the input stream with index i is now

```
struct vector {
  float x, y, z;
  vector& operator+=(vector const& v) {
    x += v.x; y += v.y; z += v.z;
    return *this;;
  }
};

vector f(vector x1, float m1, vector x2, float m2) {
  return vector ... // terms omitted
```

```
}

void nbody(vector x[N], float m[N], float F[N])
{
  for(int i = 0; i < N; i++) {
   vector f_acc = vector();
   for(int j = 0; j < N; j++) {
    f_acc += (i == j) ? vector() : f(x[i],
          m[i], x[j], m[j]);
   }
  F[i] = f_acc;
  }
}
```

Listing 3.10 N-body calculation in C++ with nested loops.

```
void nbody_tiled(vector x[N], float m[N],
  float F[N]) {
  // loop over tiles
  for(int tile = 0; tile < N/1; tile++){
   vector f_acc[1] = {vector()};
   // loop over 2nd bodies
   for(int j = 0; j < N; j++){
    // loop within tile
    for(int offset = 0; offset < 1; offset++) {
     int i = tile * 1 + offset;
     f_acc[offset] += (i == j) ? vector() :
           f(x[i], m[i], x[j], m[j]);
    }
   }

   // assign results of tile
   for(int offset = 0; offset < 1; offset++) {
    F[tile * 1 + offset] = f_acc[offset];
   }
  }
}
```

Listing 3.11 N-body calculation in C++ after loop tiling.

split up into tiles that are internally addressed by an offset. By adjusting the size of the tiles *l*, we can then implement a processing pipeline in hardware with correct and resource-efficient static scheduling. The tiled input stream allows the processing of *l* independent reductions in the hardware loop. In software, an auxiliary loop at the end of nbody_tiled() is needed to store the results of a tile. In hardware, this loop is not needed and the tile is simply forwarded to the next pipeline or the memory controller.

The resulting hardware pipeline can be seen in Figure 3.10. Like the C++ program, it uses the variables *i, j, offset,* and *tile* in the same sequence from a set of chained counters (not shown). The pipeline is fed by two input streams, e.g. from the memory controller, that provide the position and mass of the bodies. The first input receives every pair $(\vec{x}, m)$ once from memory, but *l* pairs at a time when *offset* and *j* are both zero. The pairs are separated by a multiplexer and fed into $\vec{f}$. The second input reads the data *N* times in single pairs of $(\vec{x}, m)$ when a calculation for a new tile is started (*offset* = 0). Otherwise, the former values are repeated by the register in the inputs. This way, the accumulator can compute *l* independent sums for a set of *l* forces and send the results to the output when done. Besides the controlled inputs and outputs, the extra logic needed for-loop tiling is a multiplexer that selects the required data item from a tile and a more complicated set of counters for the control logic.

Note that for-loop tiling, the input length *N* must be an integer multiple of the loop latency *l*. This property can either be satisfied by extending the input with null data, or by increasing the latency of the loop with delay registers. The introduction of delay registers is usually the preferred option since the throughput of the pipeline is maintained and spare registers are often abundant on an FPGA. The process is known as C-slow [77].

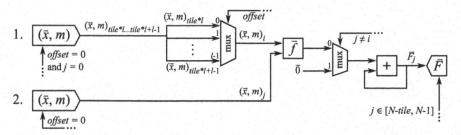

Figure 3.10   Pipeline of the n-body force calculation with loop tiling for a floating-point accumulator with latency *l*. The chained counters that generate the values of *j, tile*, and *offset* have been omitted for clarity and follow Listing 3.11. Boxed operators have a latency greater than one.

Instead of loop tiling, the data could have been reduced in memory as an alternative. In particular, the forces could have been accumulated in the DRAM, and no feedback loop in the FPGA would have been needed. Each step in the calculation would have read properties of both bodies involved, calculated the force between them, and updated the total force on the first body. However, the increments of the force at every clock cycles would have put more load on the memory interface. Loop tiling gives the hardware designer an additional design pattern that requires only one write operation for the reduced value, i.e. the total force on a body.

In general, loop tiling can always be applied for nested loops with any reduction at the innermost level where the dependencies allow the interchange of both loops. After applying loop tiling, the application can be speeded up further using the techniques described above, such as loop parallelization and (partial) loop unrolling.

3.3.3.5 Loop interweaving

Loop tiling requires at least two nested loops. If only one loop is present with a body that cannot be reduced to have a latency of only a single clock cycle, and the loop count is variable or large, none of the described techniques can be implemented and expensive additional hardware resources may be needed to counter the effects of l split substreams in the loop as described for a generic floating-point accumulator [74]. It can be advisable to search for a change in the problem domain of the application that is to be ported to allow multiple independent input streams to be interwoven. Gustafson's law states that the user is usually interested in processing more data sets in the same time [33], which can, but does not have to require the acceleration of the individual calculation. Often, the original problem can benefit from or tolerate the parallel execution of l independent calculation. Hence, while technically very similar to loop tiling, loop interweaving was chosen as a different term to describe a change induced by hardware constraints in the problem domain.

As an example, the pseudo-random generator for a constant bit width of the Java programming language [78] in Listing 3.12 can be modified to produce l interweaved substreams of random numbers. It is based on the linear congruential method [79]. When implemented similarly as drafted in Figure 3.9, it will produce a different stream of random numbers that are likely not to violate the constraints of the consumer of these numbers.

3.3.3.6 Finite-state machines

As a last resort, loops that can be computed by general-purpose CPU can always be implemented as one or multiple finite-state machines (FSM). Since

```
private long seed;
protected  synchronized  int next(int bits)
{
  seed = (seed * 0x5DEECE66DL + 0xBL)
      & ((1L << 48) - 1);
  return (int) (seed >>> (48 - bits));
}
```

Listing 3.12 The random number generation in Java qualifies for-loop interweaving.

FSMs rely on a sequential model of execution, any MISD parallelism has to be translated to a series of states. In the most extreme cases, the resulting system resembles a (small) CPU. Under the limitations of an FPGA, such as a clock frequency that is lower by an order of magnitude, FSMs do only qualify for application acceleration where they form a small part of the total design and transferring the data to the host CPU for computation would be considered too expensive. In the designs presented in this book, FSMs were chosen for control logic with nested loops that have a small resource footprint, but would require a complicated static dataflow design, and that would obstruct the actual dataflow logic with control logic.

3.3.4 Summary

Loops supposedly constitute the building blocks of sequential computing that need the most effort when porting them to an efficient hardware design. Unrolled, parallelized, and cascaded loops can be freely combined with each other and the pipelines formed by sequences and conditionals, as they all produce one data item per clock cycles. Tiled loops, however, have a different input and output pattern and require buffer FIFOs in between them for a free combination with other loops or acyclic pipelines.

The loops presented in software have all been for-loops. The choice was motivated by the linear access of data elements that is simplified by the index variable. If the access pattern of the input data does not advance linearly, the data must be reordered before. Some examples have been given in Section 3.1 and require the support of the memory controller.

Nested loops in software that do not qualify for-loop tiling or one of the other dataflow techniques can be rewritten into a single large loop with additional control variables that govern the execution of the inner loop's bodies. If the resources needed for these control loops exceed the FPGA budget, they can be moved to a slower FSM with resource sharing.

Since loops in software impact hardware design decision on a wider part of the design space than (small) conditionals and sequences, the identification of loops in a software program must be part of the hardware design phase right after the computational kernels have been isolated. The educational bottom-up presentation in this section hence is opposed to the top-down approach needed during the translation process.

3.4 Efficient Bit and Number Manipulations

A high-performance application can choose from multiple native encodings for numeric values on CPUs. Signed or unsigned integer encodings provide exact representations of the value and store numbers, depending on the required number range, in 8-, 16-, 32-, or 64-bit words. Numeric values with a much higher range can be encoded as floating-point (FP) numbers that span many orders of magnitude, but limit the precision to a fixed number of significant bits. The CPU hardware, namely the ALU and the FP unit, is optimized for processing numbers with these encodings in hardware. Other encoding must be emulated and is therefore used only cautiously by programmers.

In reconfigurable computing, all numeric operations are implemented with lookup tables and flip-flops, with the single exception of DSPs that reduce the resource footprint of multiplications. Hence, the encoding of numeric values is not predetermined by the hardware. Contrary to software applications, the most efficient design has the bit width of each stream of numbers reduced according to the minimal required precision and range. It is the task of the hardware designer to assess the required computing precision from the precision of the input values and the wanted precision of the output through error propagation or in simulation.

3.4.1 Encoding

The encoding of numbers has influence of the resources required to carry out arithmetic operations on the FPGA, but also on the amount of storage required and, as a consequence, the bandwidth in numbers per time that can be retrieved, processed, and stored. An overview about the resources needed to perform basic arithmetic is shown in Figure 3.11 for signed integers with 32 and 64 bits and floating-point representations with single and double precision. Single-precision number representations have the same bit width as 32-bit integers and therefore the same amount of information. The same applies to the floating-point representation with double precision and 64-bit integers.

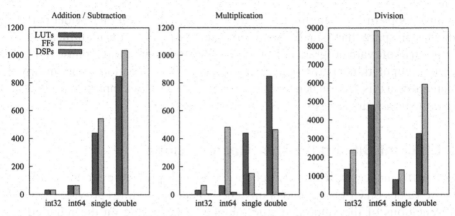

Figure 3.11   Resource usage for basic arithmetic with different data types measured on a Xilinx Virtex-6 with MaxCompiler. Addition and subtraction are cheap with integers and expensive with floating-point representations. Multiplication requires more LUTs for floating point, and division is by far the most expensive operation (note the change in scale). Virtex-6 FPGAs contain two FFs for every LUT on each logic slice.

However, the number range, the precision, and the resources needed for basic arithmetic operations vary vastly.

3.4.1.1 Integers and fixed-point representations

Unsigned integers are the natural representation for pixel values in a rasterized image. The width of an integer for color depth in popular image formats is defined as multiples of a byte, but an FPGA implementation can benefit from adjusting the width to the actual information content. Cameras for microscopy, for example, encode the intensity as a 16-bit integer, but commonly do provide only up to 12 bits of resolution. The extra four bits are zeroed and can be omitted without losing information. As indicated in Figure 3.11, reducing the bit width also reduced the FPGA resources needed for adding and subtracting by a linear factor, and by even more for multiplication and division [80].

The encoding of unsigned integers follows their representation in the binary system. For a bit vector of n bits, the range of values that can be stored is $[0, 2^n)$. For signed integers, modern FPGAs support the two's complement encoding best. It represents a negative number with absolute value a as the same bits like an unsigned integer $\bar{a} + 1$ and encodes a range of $[-2^{n-1}, 2^{n-1} - 1)$. The encoding allows integers to be added and subtracted with the same hardware as unsigned integers. Addition and subtraction are equally expensive, as subtraction can be reduced to addition by negating the subtrahend arithmetically first. The logic needed for negation consists of a binary negation and an increment by one and can be absorbed by the lookup tables of the adder without occupying additional resources (Figure 3.11).

To store fractional values, the integer representation can be extended naturally by adding additional bits after its least significant bit (LSB). The resulting fixed-point format with a fractional part of f bits can then hold multiples of 2^{-f} exactly and, depending on the number of fractional bits, approach real numbers with arbitrary precision. The bit layout is shown in Figure 3.12(a). The value of a signed number encoding with n integer and f fractional bits $a_{n-1}a_{n-2}\ldots a_0, a_{-1}a_{-f}$ is

$$v = -a_{n-1}2^{n-1} + \sum_{i=-f}^{n-2} a_i 2^i \tag{3.6}$$

Besides integers and fixed-point number representations, a third class can be constructed by removing the lower bits of an integer. For values that are known to contain a fixed level of noise, the lower bits do not have to be stored or processed and can be omitted. The logic is similar as for fixed-point representations, but with a negative number of fractional bits f (Equation 3.6).

Fractional numbers are cheap to add, subtract, and multiply when compared to floating-point number representations, given that the range of values allows an efficient encoding. The encoding of numbers also affects all other arithmetic and logical operations. Integer and fractional encodings of numbers

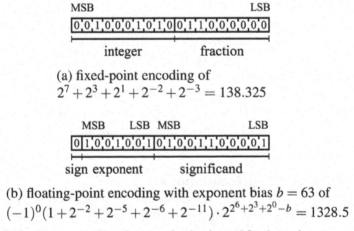

(a) fixed-point encoding of
$$2^7 + 2^3 + 2^1 + 2^{-2} + 2^{-3} = 138.325$$

(b) floating-point encoding with exponent bias $b = 63$ of
$$(-1)^0(1 + 2^{-2} + 2^{-5} + 2^{-6} + 2^{-11}) \cdot 2^{2^6 + 2^3 + 2^0 - b} = 1328.5$$

Figure 3.12 Binary encoding of custom fixed-point and floating-point representations.

can carry out the following operations especially well. Compared to the von-Neumann architecture, where each of them requires a full cycle, these functions can often be merged with other functions without affecting the timing of the FPGA design. An exhaustive overview about efficient bit manipulations of numbers in integer and fixed-point representations can be found in Hacker's Delight [72].

- **Shift:** Logical shifts and rotations by a constant amount require no FPGA resources and can be implemented by wiring alone. Arithmetic shifts that preserve the sign are also very cheap with few lookup tables.
- **Logic functions:** The binary functions "and," "or," "not," and related can be implemented in few lookup tables. They operate locally only and do not require a carry chain.
- **Additions and subtractions:** Modern FPGAs contain special logic to implement carry chains for adders and subtracters. When compared to floating point, very few logic resources are required.
- **Multiplication:** When a number is multiplied by a constant in fixed-point representation, the multiplication can be broken down into a number of additions and shifts, one for each set bit in the binary encoding of the constant. If the constant is a power of two, the multiplication becomes a single-shift operation. For general multiplication, modern FPGAs offer DSPs to decrease the number of flip-flops and lookup tables.
- **Truncation:** To avoid the introduction of a bias, fixed-point numbers must be rounded properly before truncation. Rounding half to even avoids a rounding bias and requires an adder to support rounding up, which again is cheap for fixed-point representations.

The canonical example of image processing, the convolution with a constant kernel, can be done efficiently in fixed-point encoding as the fractional part of the kernel matrix can be described with few extra fractional bits for common image manipulations. The multiplications by a constant can be split into adders and shifts, and the adders are supported well by modern FPGAs. Hence, for image processing, fixed-point representations are often sufficient. Only where the values of variables span many orders of magnitude floating-point encodings provide better resource utilization.

3.4.1.2 Floating-point representations

The encoding of a custom floating-point format is shown in Figure 3.12(b). Numbers in floating-point representation are stored by splitting the normalized form $(-1)^s \cdot 1.T \cdot 2^E$ of a number into its significant bits T with sign s and an exponent E to the power of two for scaling. Like fixed-point encodings, floating-point encodings rely on a finite number of bits and can only store an approximation of all but a finite set of real values within their range. Whereas the maximum encoding error of a fixed-point encoding is the same for all values, the error of a floating-point encoding depends on the value of the exponent. Since many measurement processes produce a relative error, a floating-point encoding can store the results efficiently at the expense of a more complicated implementation of the arithmetic operations.

The arrangement of bits is ordered by significance and its parts are encoded in a way to allow comparison of numbers with the same hardware that is also used to compare integers. The first bit stores the sign of the encoded value. The exponent is stored next with a bias instead of two's complement encoding to make its range non-negative. For an exponent with w bits, the value of the bias is $b = 2^{w-1} - 1$, resulting in a symmetric range for the exponent. Finally, the bits of the absolute value of the significand follow minus the first bit, which is known to be always one for normalized floating-point representation.

In the IEEE standard 754 for floating-point arithmetic [81], only certain lengths of significands and exponents are defined, with single (binary32) and double (binary64) precision being the formats most widely supported by CPUs and the only formats supported by graphics cards. The IEEE standard also defines a number of special values that encode ± 0, subnormal numbers, $\pm \infty$, and results of invalid operations (not a number – NaN). If these special values are known to not occur in an operation, the corresponding hardware support can be removed from the FPGA to save resources. Moreover, the format of the encoding can be changed at will. To change the range, the number of bits for the exponent can be modified, and a change in the number of bits for the significant alters the relative precision of the numbers. The standard defines the values of and operations on numbers encoded as floating point according to the numbers of bits for significant and exponent and can therefore naturally be extended to also apply to custom floating-point encodings (Table 3.1).

During the development of an algorithm, when the correctness of the program is still unproven, double precision is the most convenient way to avoid overflows and incorrect results from insufficient accuracy. Afterward, when speed becomes a concern, the number of bits for significand and exponent can be reduced, or the algorithm may even be changed to calculate with fixed-point encodings. For multiplications and divisions, the reward in chip area and therefore in power reduction is more than linear. On a Xilinx Virtex-6, the resource usage for lookup tables and flip-flops almost drops by a factor of

Table 3.1   Floating-point encoding according to IEEE standard 754 [81]. Exponent bias $b = 2^{w1}$, sign s. Single precision: $(w, t) = (8, 23)$, double precision: $(w, t) = (11, 52)$

Exponent E (w bits)	Significand T (t bits)	Value
$\in [1.2^w - 2]$		Normal number, $(-1)^s \cdot 2^{E-b} \cdot (1+2^{2-t} \cdot T)$
$= 0$	$= 0$	$(-1)^s \cdot 0$
$= 0$	$\neq 0$	Subnormal number, $(-1)^s \cdot 2^{\text{emin}} \cdot$ $(0 + 2^{2-t} \cdot T)$, with emin $= 2 - 2^{w-1}$
$= 2^w - 1$	$= 0$	$(-1)^s \cdot \infty$
$= 2^w - 1$	$\neq 0$	NaN

three for both addition and multiplication when moving from double to single precision. For divisions and square roots, the ratio is even better and drops by about a factor of four [82]. With FPGAs, every stream of numbers can be adjusted on the bit level to only use the minimal amount of logic for precision, range, and the special floating-point values shown in Table 3.1.

3.4.1.3 Alternative encodings

Besides floating-point and fixed-point encodings, other encodings for special demands exist. When processing numbers within a huge range, the logarithm of a number can be stored in a fixed-point format, similar to a floating-point encoding without exponent. The encoding of a value v becomes the fixed-point encoding of $\log_2 v$. Multiplication, division, and exponentiation are mapped to addition, subtraction, and multiplication and become much cheaper (Equation 3.7). However, addition and subtraction in a logarithmic number system are resource-intensive. The encoding is commonly used for digital signal processing with high dynamic range [83].

$$\log_2(v \cdot w) = \log_2(v) + \log_2(w)$$
$$\log_2(v/w) = \log_2(v) - \log_2(w) \qquad (3.7)$$
$$\log_2(v^w) = w \log_2(v)$$

For image processing, the range of pixel values is limited and can be stored as an unsigned integer. Still, space can be saved by not storing the noise of an image. The number of photons N captured by a perfect sensor follows the Poisson distribution, where the noise has a width of $\sigma = \sqrt{N}$ [84]. Since the square root of a number consists of half as many bits as the number itself, half the significant bits of the integer encoding contains noise and can be omitted. To do so, the square root of every intensity is taken before it is stored [85]. For retrieving, the intensity is squared again. This saves half the bits required to store the maximum intensity and can be used for compressing images. The error Δg introduced by rounding the square root to the nearest integer with error $\Delta \delta$ between rooting and squaring is

$$g(N) = \left(\sqrt{N + \delta}\right)^2 \qquad \bar{\delta} = 0 \tag{3.8}$$

$$\Delta g = \sqrt{\left(\frac{\partial g}{\partial \delta}\right)^2 \Delta \delta^2} \qquad \Delta \delta^2 = \int_{-0.5}^{0.5} \left(\delta - \bar{\delta}\right)^2 \mathrm{d}\delta = \frac{1}{12} \tag{3.9}$$

$$= 2 \left(\sqrt{N} + \bar{\delta}\right) \sqrt{\frac{1}{12}} \tag{3.10}$$

$$= \sqrt{\frac{N}{3}} < \sigma \tag{3.11}$$

The lost information Δg therefore only causes an additional error smaller than the value of the lower half of the significant bits of N. A 12-bit signal from an optical camera can be compressed to a 6-bit word. Compared to simply truncating the lower six bits, extracting the root consumes more chip resources, but preserves most of the information for small intensities. For multiplication, division, and exponentiation, the numbers can be processed directly the same way as integers and do not have to be squared again.

Besides the presented encodings, a plethora of number and object formats exist. In particular, for larger data structures such as images, movies, or matrices, data can be compressed until it reaches a maximum density of entropy. Run length and difference encoding can be implemented well with FPGAs, and also, the lossy compression standards JPEG for images and MPEG for movies have been successfully ported to FPGAs [86, 87].

3.4.2 Dimensioning

The numeric data that arrive at the input of a hardware pipeline come with a range and a precision. Both properties do not have to exploit the maximum number of bits of their encoding. It is, on the contrary, rather common to receive a stream of values encoded as floating point with double precision where the distribution of values makes use of only a small fraction of the exponent and the significand. CPUs only support a limited set of number formats, and using a format not natively supported can lead to a performance penalty. On the FPGA, however, every saved bit will decrease the logic area. It is therefore advisable to trace the origin of the input data and reduce the bit width of the input values according to their true information content at the beginning of the pipeline, and further process only the minimal number of bits needed in the following stages. The resulting hardware will then consume less resources and will better meet timing requirements.

In the hardware pipeline itself, the range and precision grows or shrinks with each operation. In the following sections, the required adjustments to the encoding are examined to avoid a loss in precision due to overflows or truncation.

3.4.2.1 Range

When the range of a value exceeds the capacity of a number encoding, a positive or negative overflow happens and the value is lost in its entirety. Even with floating-point encodings that represent these cases as plus or minus infinity, the value is trapped and will not approximate a real value again when applying basic arithmetic. It is therefore of importance to provide enough bits for the integer part of a fixed-point representation or the exponent of a floating-point representation.

For signed fixed-point encodings with n integer bits, the range of possible values covers the values in $[-2^{n-1}, 2^{n-1})$ and changes with each operation. The adjustment of the integer bits can be partially automated through a technique known as bit growth. It adjusts the number of integer bits for each operation to avoid underflows. On the level of an individual operation, the bit growth for fixed-point encodings is shown in Table 3.2. For additions and subtractions, this is a single bit to avoid overflows. For multiplications and divisions, the integer part of a fixed-point representation must grow by a larger number of bits, depending on the number of bits of its arguments.

The bit growth of the integer part in Table 3.2 is given for the worst case, where the arguments occupy the full range of possible values. For chains of additions or subtractions with input values $a_1, a_2, \ldots, a_m$ with n_i integer bits, the number of bits n' for the integer part of the result only grows logarithmically, since the order of magnitude rises with the binary logarithm

Table 3.2 Maximum bit growth of basic operations for arguments a_1, a_2 with integer bits n_i and fractional bits fi on signed fractional numbers. The results of divisions and square roots cannot be computed in general without losing precision

Operation	Resulting Integer Bits	Resulting Fractional Bits
Addition	$\max(n_1, n_2) + 1$	$\max(f_1, f_2)$
Subtraction	$\max(n_1, n_2) + 1$	$\max(f_1, f_2)$
Multiplication	$n_1 + n_2$	$f_1 + f_2$
Division	$n_1 + f_2$	∞
Square root	$[n_1/2]$	∞

when adding the maximum value repeatedly (Equation 3.12). The rule also applies when dimensioning accumulators for m summations.

$$n' = \max(n_1, n_2, \ldots, n_m) + \lceil \log_2(m) \rceil \qquad (3.12)$$

For multiplications and divisions, a similar rule does not exist. However, multiplications by a constant can, depending of their value, assign zeros to the lower bits of the fixed-point representation. If the constant factor happens to be a power of two, the multiplication or division degrades to a arithmetic shift operation, and as a consequence, the bit size needed for integer and fractional part is kept constant. For multiplications with a constant that is represented by the sum of multiple powers of two, the largest power defines the growth of the integer part, and the smallest power defines the shrink of the fractional part.

Extraneous bits in fixed-point numbers can also be identified by analyzing the dataflow backward, removing all bits that do not influence the result of the program. For example, a user may only be interested in the modulus of a number by a power of two, making the information carried by the higher bits redundant. An automatic forward and backward analysis of unused bits can be carried out automatically [88], but is not widely available yet.

For floating point, the range of numbers that can be encoded is defined by the number of bits of the exponent. With every extra bit, the range is extended far wider than by exponential growth. For single precision, an 8-bit exponent is long enough to store values up to about $\pm 10^{38}$, and the 11-bit exponent of double precision allows values that surpass $\pm 10^{307}$. It is therefore much less likely for an overflow to occur when compared to fixed-point numbers and usually sufficient to introduce a one-bit safety margin for the exponent after profiling the data. Hence, bit growth for the floating-point exponent is rarely needed, and the hardware designer can focus on the precision of the significand.

3.4.2.2 Precision
The precision needed for a computation depends on the uncertainty of its inputs and the allowed error at its output. If only the uncertainty of its inputs is taken into account and the highest accuracy is required for the output values, the number of fractional bits can be derived from Table 3.2 for fixed-point numbers. However, division, square root, and other operations produce values that cannot be encoded exactly with fixed-point numbers, such as 1/3 or $\sqrt{2}$. These results must be truncated after rounding.

Rounding numbers must not introduce a bias, especially in iterative applications where a bias could add up and affect more than the least-significant bit. When rounding a fixed-point number with a fractional part of 0.5 to an integer, the number is therefore best rounded toward neither plus or minus infinity, but to the nearest even or odd integer. Though rounding half to even still introduces a bias toward infinity for even numbers and a bias toward minus infinity for odd numbers, the average of the distribution of results can be expected to be the same as for accurate calculations. Round to even is therefore also the standard mode for the significand of IEEE floating-point calculations. The error introduced by rounding and truncating a value toward a fixed-point encoding with f fractional bits is

$$\Delta_{\text{round}}^2 = \int_{-2^{-f-1}}^{2^{-f-1}} \delta^2 \mathrm{d}\delta = \frac{1}{12} 2^{-3f} \tag{3.13}$$

The rounding error is added quadratically to the error that is propagated through the inputs. All errors are therefore shown to the power of two. The result of a function g with arguments $x_1, x_2, \ldots x_m$ and rounding, such as a multiplication without bit growth for the fractional values, will carry an error of Ay as shown in Equation (3.16).

$$y = g(x_1, x_2, \ldots, x_m) \tag{3.14}$$

$$\Delta_{\text{prop}}^2 = \sum_{i=1}^{m} \left(\frac{\partial g}{\partial x_i} \Delta x_i \right)^2 \tag{3.15}$$

$$\Delta y^2 = \Delta_{\text{prop}}^2 + \Delta_{\text{round}}^2 \tag{3.16}$$

As a rule of thumb, the smaller error can be ignored if its absolute value is smaller by a factor of 10. The quadratic addition of errors ensures that the total error is only affected and increased by 1% by the smaller error. Therefore, a design should first focus on the errors introduced by the confidence of the input values to dimension the bit width of all numbers with a safety margin of about an order of magnitude, and only be reduced to the minimal amount of bits through simulation and deductive error propagation after the correctness of the implementation was shown.

For addition and subtraction, error propagation (Equation 3.16) adds the absolute errors of the operands quadratically. For multiplication and division, the relative errors of the results are added to get the relative error of the result. Hence, fixed-point number encodings support the error characteristic for values that are dominated by a constant absolute error. For values that have an error best described with a constant relative error, floating-point numbers are a more natural representation for the uncertainty.

A normal floating-point number with t significant bits has a precision limited to f bits, and its relative error $^A X^X$ is similar to the absolute error of a fractional number in Equation (3.13).

$$\frac{\Delta x^2}{x^2} = \frac{1}{12} 2^{-3t} \tag{3.17}$$

The number of fractional bits of a floating-point encoding defines the relative accuracy of a number. It allows for a wider variation of the number of bits during resource optimization than the number of bits for the exponent, where two bits separate the occurrence of value-destroying overflows to a waste of bits.

In summary, choosing an encoding should start with the range of the values first: Very large ranges cannot be expressed well with fixed-point encodings. For image processing, sensor data from image sensors are generally restricted to 12 or at most 16 bits and very suitable for fixed-point encodings. Operations that increase the range of the result by more than a constant number of bits, such as multiplication, division, or exponentiation, may require a cast from fixed point to floating point. Once the design avoids overflows, the precision can be reduced such that the accuracy of the output is still maintained.

3.5 Customizing Memory Access

Configurable hardware cannot only be customized in terms of computational operations, but also how it accesses data stored in memory. The performance of an algorithm that suffers from a memory bottleneck on a CPU system can be improved on an FPGA if the access patterns of the dataflow from and to memory can be adjusted to leverage its parallel capabilities. After profiling and determining which memory operations hold back the performance of the CPU implementation, the following options are available to manage state on an FPGA system.

- **Registers:** The flip-flops embedded in every logic cell of an FPGA can be combined to form registers that hold words of information. Like CPU registers, their content can be read or written at every clock cycle. Moreover, they can be accessed all in parallel and read and written at the same time, increasing the number of simultaneous register transfers immensely. Available FPGAs offer up to half a million of logic slices with eight flip-flops each, resulting in more than 60,000 64-bit registers. Although most of them are used as pipeline registers for dataflow computing, registers are still abundant on an FPGA when compared to the 16 general-purpose CPU registers of the AMD64 architecture [89]. Multiple registers can be combined with multiplexers to form small areas of addressable memory at the position where the data are needed and can be embedded into a seamless dataflow pipeline.

- **BRAM:** For larger amount of data up to the order of a megabyte, most FPGAs include BRAM. It is implemented on its own and distributed in portions of multiple kilobytes on the chip to save resources, especially flip-flops. Like memory build on flip-flops, BRAM can be read and written word by word at the same time at two independent addresses. It offers an efficient way to implement on-chip caches and dynamic stream navigation within a sliding stream window.

- **On-board DRAM:** For large amounts of memory with multiple giga-bytes of data, DRAM memory can be added on-board. The data are accessed through an instantiated memory controller on the FPGA. For maximum throughput, DRAM must be operated in burst mode, with a burst consisting of at least 4 words [90]. Also, the address pattern must be known in advance to mitigate the read and write latencies. Unlike BRAM, DRAM does not decrease the maximum clock frequency when its size is increased.

- **Host memory:** For FPGAs that are connected with a host computer, typically through the PCI Express 2.0 bus, additional memory can be provided as shared memory within the address space of the host system. The additional bus increases the latency and decreases the throughput, but can conveniently be used to exchange information between host system and FPGA card.

The FPGA memory infrastructure makes copying data cheap because it can be parallelized massively on the register and BRAM level. If memory access to external memory is needed, the algorithm should be rewritten to follow an access pattern on the data that is known in advance and can therefore be computed earlier by at least the combined latency time of the DRAM and memory controller. Since this requirement often implicates a major remodeling of the dataflow of the algorithm, a C model should be created for verification. It can afterward act as a semantic bridge between the initial software application and the hardware port.

3.5.1 Memory Layout and Access Patterns

In C, C++, and related languages, state is stored on either the stack or the heap. Access to the stack is limited to the topmost stack frame of the current function. On the heap, memory is abstracted as an almost infinite, linear memory space where chunks of memory for structs and objects can be obtained at runtime through a memory allocator. We have already seen that a stack can be removed by inlining and moving all variables to the heap, or, for recursive functions, using an explicit auxiliary stack on the heap. It comes with the advantage that all stack frames are of the same size and type for most kinds of recursion, simplifying custom memory access. Since the current stack frame moves only in steps of one unit, it can be cached in BRAM when it must be stored in DRAM due to its size.

Variables stored outside of recursive functions can be absorbed by the pipeline as described in Section 3.3. For large objects, however, the pipeline can become too wide, and the data are better stored in addressable memory to save resources. In the following sections, the difference between arbitrary addressable on-chip memory and burst-oriented DRAM will be described with more detail, as well as the implications when porting software to hardware.

3.5.2 On-Chip Memory

Memory on the FPGA that exceeds a few bytes can be implemented with BRAM and addressed with every clock cycle. BRAM offers two ports, so it can be read and written at the same time. This feature makes it blend in well into a statically scheduled pipeline, where it may consume and produce one word per clock cycle. In image processing, BRAM is used as line buffer (Figure 3.13) or to store any other data temporally. On-chip memory enables to store a halo of the current stream position.

In software programs where an operation is executed during the traversal of a data set, BRAM can be used to store data that are local to the current iteration. Doing so, convolutions on data can be generalized and implemented efficiently with a dataflow design on an FPGA. For the background calculation in localization microscopy, an entire background map for the current picture is kept in BRAM (see Section 4.1.5).

BRAM allows for the addressing of individual words and can therefore also be used where the address offset to the current stream position is dynamic. An example can be seen in the accumulator in Section 4.2.4 that uses BRAM to build a projection of a 3D volume with only weak guaranties regarding which individual accumulator will be increased next.

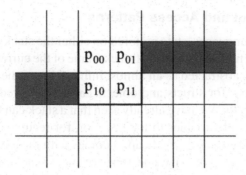

(a) Image convolution with buffered pixels.

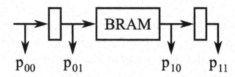

(b) Hardware implementation with BRAM and
two delay registers (address counter omitted).

Figure 3.13   Line buffering. The gray pixels must be kept in the line buffer for convolution in x-direction. The buffer can be integrated into a statically scheduled pipeline using dual-port BRAM.

In software, it can be difficult to immediately decide whether BRAM should be used. Indicators are the usage of small arrays of a few kilobytes that are accessed randomly, or limited offsets to the current index position in a for-loop.

3.5.3 Off-Chip Memory

DRAM storage on the FPGA board is organized in rows and columns, similar to a matrix. To access a byte, the row must first be selected by the controller before the data can be read in a second step. Both operations have a latency of at least 10 ns [91]. The memory controller can pipeline these operations to a certain degree, but a random memory access still consumes at least 7 ns, limiting the access frequency for random reads to less than 143 MHz. For sequential access, much higher transfer rates than 100 MB/s are reached. Here, data are read in bursts, where the selected row is accessed in large chunks of at least four words. In theory, up to 17 GB/s of data can sequentially be read or written per module.

We have seen in Section 3.1 that algorithms that can be reduced to list processing are especially well-suited for dataflow implementations. In general, algorithms that traverse data linearly can be implemented with maximum throughput from and to external DRAM. For applications that do not, data access should be re-arranged if possible.

Matrix transposition, for example, requires line-wise read access and column-wise write access at the same time. However, if the storage format is changed from a simple line-by-line format toward a collection of submatrices, with a submatrix having the size of one or multiple DRAM bursts, both operations become compatible with the coarse-grained access to DRAM. The submatrices are then transposed in parallel by wiring, and the address generator transposes the arrangement of the submatrices. For the acceleration of 3D electron tomography, the algorithm was changed to facilitate linear BRAM access for the 3D volume it reconstructs.

The addressing of DRAM in units of a burst puts an additional constraint on programs that handle data. On the CPU, the cache hierarchy and the memory controller hide the characteristics of DRAM. However, even on a CPU, truly random access to the main memory will mostly miss the cache and limit the performance by more than an order of magnitude when compared to a program that hits the cache every time (Figure 2.5). It is therefore advisable to avoid random DRAM access on CPU systems as well.

For data structures that are more complicated, such as maps or graphs, internal and external memory can be mixed. Meta information such as the management structures of a memory allocator can be stored in BRAM, where it is available with low latency and can therefore be modified much faster and, if needed, in parallel. The actual content remains on external memory in chunks of multiples of a DRAM burst, similar to the various caching systems on a CPU system.

3.6 Summary

In this chapter, the transition from control flow code to a dataflow description was presented. Due to the different nature of both descriptions, an understanding of the algorithm that is to be ported is crucial on all levels. To achieve a maximum acceleration, not only must the building blocks of a sequential program be mapped to pipelined hardware, but also the higher levels. The intended number range and accuracy must be understood to save resources, and the hot spots of the algorithms are to be identified to partition the application between dataflow hardware and the parts that stay on a CPU.

The wanted gain in execution speed can then stem from the following changes to the program.

- **Massive parallelism:** Custom hardware allows for SIMD as well as for MISD parallelism and combinations of both. In the best case, all stages of the resulting system of pipelines process data at each clock cycle, and little resources are occupied by auxiliary task such as control logic and caching. The transfer of values in registers and values can be parallelized massively.

- **Simplified hardware:** On a CPU, every piece of hardware is built for the most general use. Arithmetic units are designed to calculate on the full range of random inputs for the given and fixed encoding, and the CPU is expected to perform well on arbitrary programs. For custom hardware, the program is known in advance, and all logic that is not needed can be omitted, giving space to increased parallelism. The saved logic can be found on all levels, from saved caches down to constant propagation for basic arithmetic and logic units.

- **Reduced level of abstraction:** In custom hardware, abstractions such as virtual memory, function calls, call indirection, and the entire software stack including the operating system are removed in for a pipelined hardware design. The missing instructions save further hardware resources for the actual calculation. They must be compensated by the high-level development tools to not impair development times and the flexibility for future adjustment.

Reconfigurable hardware benefits from these advantages, but is also impacted by a higher silicon footprint for single gates and a clock frequency. The next chapter will present two example applications from the research field of biomedical image processing and reconstruction to research the true potential of reconfigurable computing in these areas.

4

Biomedical Image Processing and Reconstruction

In this chapter, two applications and their implementation with the MaxCompiler library on reconfigurable hardware will be presented. The first one, image processing for localization microscopy, was fully ported, and its accelerated algorithms were in use at several research sites at the time of the writing. The second, image reconstruction for electron tomography, was also implemented on reconfigurable hardware and is shown to be faster than its counterpart on a state-of-the-art graphics card.

4.1 Localization Microscopy

Small objects like biological cells can be made visible to the human eye with a light microscope. It produces images of high contrast in the spectrum of visible light, and the result can be seen immediately without further processing. For data storage, good optical cameras are available. Even live cells can be observed and multiple techniques exist to dye otherwise invisible structures. The limit of conventional light microscopy is reached when the structures become smaller than the wavelength of the light source. Due to physical limits, a point in the object is mapped to a blurred spot in the image by the microscope, and spots close by cannot be distinguished any more from each other. For visible light, the limit is above 200 nm, and hence, the resolution is too coarse to image nanostructures in biological cells that extend to few nanometers.

Localization microscopy can separate these close-by spots optically to a certain degree and increase the resolution by a factor of ten, depending on the quality of the recording. Localization microscopy relies on the preparation of the object with fluorophores that switch stochastically between different optical states over time when illuminated with laser light. The different states, e.g. bright and dark, are used to distinguish close-by fluorophores. A movie

 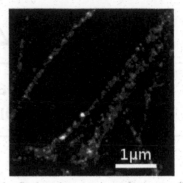

(a) conventional wide-field fluorescence microscopy   (b) localization microscopy image from a recording

Figure 4.1   Localization microscopy improves the resolution of fluorescence microscopy by about an order of magnitude. The positions of fluorescent molecules are determined with subpixel accuracy and plotted into a new image [95].

of the blinking fluorophores is recorded and the spots are later examined individually frame by frame. Typical image stacks consist of thousands of frames and span several minutes until the fluorophores are bleached out. Afterward, the center of each spot in the recording is obtained by fitting its signal distribution and the positions are plotted to create a new image of the object with superior resolution. An image that compares conventional microscopy with localization microscopy can be seen in Figure 4.1. The increased resolution allows the diagnosis of various diseases, such as the analysis and identification of specific kinds of breast cancer [92].

The data processing on the recording, including background removal, spot finding, and feature extraction, was implemented on the Maxeler DFE as part of this book. The existing algorithms were first rewritten to be feasible for dataflow computing on reconfigurable hardware and then ported and integrated into the workflow of the microscope.

4.1.1 History

The first preserved written records about optical devices capable of magnifying small objects were passed down from the ancient Romans. Lucius Annaeus Seneca described how to fill glass bowls with water to built a first predecessor of the microscope in the first century AD. The availability of glass lenses in the 16th century led to the development of microscopes with multiple lenses that provided higher magnification factors. Christian Huygens invented the first microscope with a chromatically corrected system of lenses that improved image quality further in the late 17th century when lens manufacturing advanced further.

The construction of microscope stayed a craft led by experiment and experience until Ernst Abbe and Carl Zeiss developed the physical foundations of optical microscopy that arise from the wave-like nature of light. Today, the resolution limit for classical optical microscopy is known as the Abbe diffraction limit [93], and its mathematical formulation supports the creation of microscopes that achieve a maximum magnification of about 200 nm. Many important discoveries have been made since then with the microscope as a tool, including the analysis of bacteria that cause plagues, pox, and tuberculosis or to gain insights into human cells.

To further increase our knowledge about the microscopic details of biological systems, an ever-increasing resolution was desired. Since the Abbe diffraction limit $d = \lambda/(2\,\text{N.A.})$ is proportional to the wavelength λ, ultraviolet light can be used instead of visible light to increase the optical resolution. The optical aperture N.A. of the microscope, which also defines the resolution limit, can be improved by immersing the object into oil with a high diffraction index.

The nanometer and even picometer scale became directly accessible with electron microscopes. Louis de Broglie's discovered the wave-like nature of electrons [94] with wavelengths a hundred thousand times shorter than visible light in 1924. A beam of electrons in a vacuum can be bent with magnetic and electrical fields as a substitute for lenses. Contrary to light microscopy, however, the object has to be prepared for the high energy of the electron beam and for the exposure to vacuum, making it unsuitable for live cells and affecting the geometry of biological samples. Subsequently, methods to circumvent the Abbe diffraction limit were developed to achieve higher resolutions for light microscopy.

The approach that overcame the Abbe diffraction limit first was stimulated-emission-depletion fluorescence microscopy (STED). The method was first published in 1994 [96] and requires the object to be colored with a fluorophore that acts as a marker for a certain type of protein in a cell. The fluorophores then emit light when excited by a laser of the excitation frequency, while the remaining parts of the object remain dark. STED scans the object sequentially with a higher resolution as the Abbe diffraction limit would permit. It does so by first depleting the fluorophores in the close environment of the current scan position with a doughnut-shaped laser spot, putting them into a dark (disabled) state for a short time. Immediately thereafter, the light source is switched to a second laser that excites only the fluorophores left enabled in the center of the doughnut-shaped depletion zone. Since the center can be smaller in diameter than the wavelength of light, the resolution is improved to 35–70 nm. The final image is compiled afterward from all scan positions to be visible to the human eye.

Localization microscopy requires fluorophores in the object as well. Instead of switching them off with a depletion laser, special fluorophores oscillate themselves between a dark and a bright state. Stochastic optical reconstruction microscopy (STORM) [97] uses the stochastic blinking of the fluorophores in the object to allow their separation in a recording. With the bright state being active much shorter than the dark state, the center of each spot can likely be determined individually. Multiple methods based in this principle have been developed since 2006 by several research groups [97–99].

The center of each spot from an active fluorophore can be determined with a higher accuracy than its width on the image sensor when it is not obstructed by the spots of neighboring fluorophores. The location of a fluorophore can be determined with an accuracy of up to 5 nm, and with about 10 nm on average, improving the resolution compared to conventional microscopy by at least an order of magnitude.

The process does not only require a laser microscope (Figure 4.2), but also considerable computing power to find the spots during their bright stage on each frame and to determine the center by fitting the spot. Additional tasks are background elimination and noise reduction. Depending on the number of fluorophores in the region of interest, a typical recording of five minutes with 10 frames per second can take multiple hours to process until finally the super-resolution image is plotted from the results. Handling errors

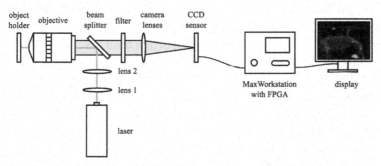

Figure 4.2   Simplified setup for localization microscopy. A laser illuminates the object and causes its fluorophores to blink stochastically. They act as optical markers and are recorded by a CCD camera. The application on the DFE determines their positions and produces a super-resolution image.

during preparation and recording can therefore not be discovered until hours have passed. The latency causes the work flow of the experimenter to be slowed down considerably, hindering the spread of the technique in the biomedical community. To achieve real-time performance, the image analysis has to be as fast as the recording. For a typical frame rate of 10 frames per second, the processing of an individual frame must therefore not exceed 100 ms.

4.1.2 Physical Principles

Even a perfect light microscope could not image a point-like light source in its focus area to a spot with zero width. Instead, the wave nature of light causes the spot in the image to be blurred, and the resolution of the microscope is limited by physical principles.

The first mathematical description of the diffraction of light that causes this effect was done by George Biddell Airy in 1835 [100]. When the light of the source travels through the optical system, it reaches the image on multiple paths. In the center of the spot, all waves interfere constructively and the signal is brightest here. When moving radially away from the center, the interference gradually turns destructive, producing a dark ring around the signal. Even farther, minor maxima can be seen. To calculate the pattern, the light has to be integrated for each possible path through the microscope along the optical axis. For systems with a circular aperture and a light source far away, the resulting pattern from the integral is described in Equation (4.1). It contains the Bessel function J_1 of the first kind and order one [101] and describes the intensity I at radius x from the center for a spot with total intensity I_0.

$$I(x) = I_0 \left(\frac{2J_1(\pi x)}{\pi x} \right)^2 \qquad J_1(y) = \frac{1}{\pi} \int_0^\pi \cos(\tau - y\sin(\tau)) \mathrm{d}\tau \qquad (4.1)$$

The intensity distribution, named Airy pattern after its discoverer, can be seen in Figure 4.3 from above and in profile. It is shown in units of λf, where λ is the wavelength and f is the focal width. The first minor maximum reaches less than 2% of the relative intensity. The main maximum can be approximated well between the zeroes of the function with a 2D Gaussian profile ($f\vec{x}$) with center (μ_x, μ_y,) width (σ_x, σ_y), and total (integrated) intensity Q. For objects with more than one point-like light source, the resulting image is obtained by folding the input with the Airy disk as the points spread function.

$$f(\vec{x}) = \frac{Q}{2\pi\sigma_x\sigma_y} e^{-\frac{1}{2}\left(\left(\frac{x-\mu_x}{\sigma_x} \right)^2 + \left(\frac{y-\mu_y}{\sigma_y} \right)^2 \right)} \qquad (4.2)$$

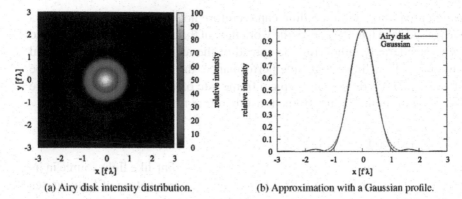

(a) Airy disk intensity distribution.   (b) Approximation with a Gaussian profile.

Figure 4.3   The profile of an Airy disk. It describes the picture of a point-like light source. The minor maxima are barely visible. The intensity distribution can be approximated well with a Gaussian fit.

The Gaussian profile is much simpler to handle than the nonclosed Bessel function in the mathematical description of the Airy profile. It is therefore preferred when fitting the spot to extract its features. Furthermore, the Gaussian profile has a finite width $\sigma_{x,y}$ that can be used after fitting to identify spots of fluorophores that have drifted out of focus and are blurred by a larger extent.

For a microscope with opening angle θ, the Abbe diffraction limit for the minimum distance d of two points embedded into a medium of optical density n that can still be distinguished is given in Equation (4.3) [94]. For commercial microscope, the limit is $d \simeq \lambda/2$. The plot in Figure 4.4 depicts that if the

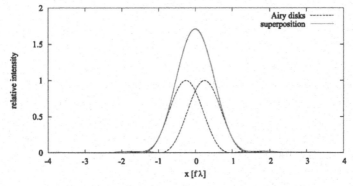

Figure 4.4   Abbe diffraction limit. The Airy disks of two fluorophores located closer than $\lambda/2$ cannot be distinguished any more when they are both in the bright state. Stochastic blinking, however, allows to fit and locate them separately.

points are closer, the Airy disks merge into a combined intensity distribution with a single maximum, and the individual locations cannot be distinguished any more from the superposition.

$$d = \frac{\lambda}{2n \sin(\theta)} \tag{4.3}$$

With localization microscopy, two spots with distance d most probably do not share the same optical state all the time during the recording, since the fluorophores are chosen to stochastically change their state many times during a recording. The activation times are chosen to be sparse, causing the visible period to be much shorter than the invisible. We can therefore fit the first spot when the second one is dark and vice versa, beating the Abbe diffraction limit.

Other fundamental limits still exist, and they constrain the newly improved resolution. First, the camera sensor is made of an array of discrete pixels that defines the location accuracy of a hit by a single photon. A fit takes many photons into account and mitigates this uncertainty. The individual error depends on the fit, but the more photons are available, the more precise a Gaussian profile can be fitted. Modern CCD cameras for localization microscopy have high photon efficiencies of more than 90% to miss as few photons as possible.

The second fundamental uncertainty is caused by the stochastic nature of photons. Any photon from a fluorophore has a small chance to hit a pixel on the camera sensor. The repeated process therefore follows a binomial distribution. For small individual probabilities, the binomial distribution approaches the Poisson distribution as a special case [84]. It is given in Equation (4.4).

$$P_\lambda(n) = \frac{\lambda^n}{n!} e^{-\lambda} \tag{4.4}$$

Its single parameter A defines the mean and the width $\sigma = \sqrt{\lambda}$. For increasing values of λ, the distribution approaches a Gaussian distribution with the same mean and width. For small λ, it becomes more skewed, but maintains zero probability for negative counts n. A plot of the distribution and a sample spot with pixelation and Poisson noise can be seen in Figure 4.5. The relationship between width and mean allows us to estimate the noise from the measured value as $\sigma_{est} = \sqrt{n}$.

The noise is further increased by the background of the image that is created by structures out of the focus plane of the microscope. Complex objects like entire cells disperse photons from its organelles and produce an inhomogeneous background image that also follows the Poisson statistic. Other sources of noise include dark noise from thermal photons, clock-induced

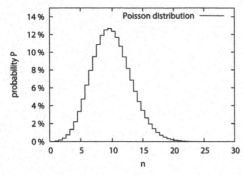 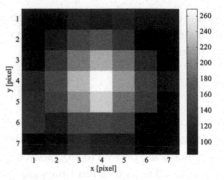

(a) Poisson distribution with mean $\lambda = 10$ and width $\sigma = \sqrt{\lambda}$

(b) Simulated Airy disk with background on a CCD sensor

Figure 4.5   Distortion of an ideal Airy disk by pixelation and Poisson noise. Both effects limit the accuracy of a Gaussian fit and decrease the confidence of the obtained localization at its center.

noise that is introduced when the pixel charges are moved to the border of the sensor for read-out, and finally readout noise that stems from the imperfect electric components before digitalization [102].

The magnitude of Poisson noise is given by the width of its distribution, and the relative width $\lambda\sqrt{\lambda} = 1/\sqrt{\lambda}$ decreases with a brighter image, resulting in better fits and smaller localization errors. To excite as many fluorophores as possible, monochromatic laser light is used that is tuned to match the fluorophores excitation frequency. Still, high intensity laser light used for illumination acts toxic and can damage or destroy a biological sample during the recording. The experimenter must therefore balance the intensity to maintain the integrity of the sample and the resolution of the localization image.

4.1.3 Localization Algorithms

The aim of this part of the book was to accelerate the image processing algorithm for localization microscopy. With a recording as an input that extends over many frames in time, the steps that have to be performed at least to create a super-resolution image from the input can be described as the following.

1. **Background removal:** This step eliminates the background of each frame caused by fluorophores out of focus and photon scattering, either on each frame individually or by taking into account information gathered from previous ones. The requirements depend on the capabilities of the next step.

2. **Spot detection and isolation:** The signal of every spot must be identified and its position must be forwarded to feature extraction. The quality of this processing step is defined by the number of false positives and negatives that are easily induced by the photon noise in the recording.
3. **Feature extraction:** The most important information is the position of the fluorophore, derived from the center of the Airy disk, and the confidence it was measured with. Other features are the discrepancies from the Airy disk profile, caused by of out-of-focus fluorophores. Such spots are to be excluded from the final super-resolution image.
4. **Generation of the super-resolution image:** Finally, the positions are to be visualized. The plot should include information about the resolution at each position. The resolution depends on the confidence of the individual positions and the density of positions in their neighborhood.

The algorithm can only be split into the presented parts for recordings with sparely distributed spot signals. Here, overlapping spots can be discarded and the quality of the final localization image is still maintained. If too many spots overlap with distances at the Abbe diffraction limit, they cannot be examined any more with individual Gaussian fits. In this case, the fitting function has to be adjusted to also support approximation of multiple centers [103]. In the extreme case, clusters of computers have to be used to calculate with a hidden Markov model for typical localization input imagery with 256×256 pixels and more than 1000 frames [104]. In the following, we will focus on sparse recordings with sporadic overlaps.

Following the consequences of Amdahl's law (Figure 2.7), as many parts of the chain of algorithms as possible should be done in parallel on the DFE. Therefore, a suitable algorithm is wanted that maintains accuracy and can be efficiently ported to a dataflow pipeline on an FPGA.

The software version that was used as a starting point and as a reference for this book was written in the programming language MATLAB [105]. Its different steps follow the outline presented above, and its algorithmic parts will be described in the following sections next to the available alternatives. As we will see, porting the algorithms to a DFE required a redesign of the program on all levels.

4.1.4 Background Removal

The level of the background and its noise level on each frame must be known in the first place for reliable spot detection. The intensity of different recordings varies with the chosen power output of the illuminating laser, and therefore, a threshold for spot detection must depend on the difference between image and background as well as the noise level. The noise level can be estimated for each pixel by calculating the square root of the photon count, since Poisson noise is the dominant source of noise in the system.

A first approach for background elimination that is implemented in the MATLAB version is to calculate the difference of each frame I_n and its previous frame I_{n-1} pixel by pixel. Spots that are only visible in I_n clearly emerge from the subtraction, and for adjacent frames, the background can be considered constant. This procedure is only recommended for backgrounds with strong inhomogeneities, since the Gaussian fit is also able to adjust to a certain background level. For the noise of the resulting image I', we find for the variance of the noise of each input image using error propagation:

$$I'_n = I_n - I_{n-1} \tag{4.5}$$

$$\sigma^2_{I'_n} = \sigma^2_n + \sigma^2_{n-1} \tag{4.6}$$

$$\sigma_{I'_n} \simeq \sqrt{2}\sigma_n \tag{4.7}$$

The noise level of the differential image is therefore increased by a factor of $\sqrt{2}$ when compared to the noise σ_n of the input images I_n. Removing the background should, however, only add as few noise as possible to not further disturb spot detection and feature extraction in the later steps. As an alternative, we can assume that photobleaching changes the background only slowly. Thus, we can subtract it with few additional noise by first calculating the sliding average of the last M frames to obtain a map $B_{n,\mathrm{SA}}$ of the background.

$$I'_n = I_n - B_{n,\mathrm{SA}} \quad B_{n,\mathrm{SA}} = \frac{1}{M} \sum_{i=1}^{M} I_{n-i} \tag{4.8}$$

$$\sigma^2_{I'_n} = \sum_{i=0}^{M} \left(\frac{\partial I'_n}{\partial I_{n-i}} \right)^2 \sigma^2_{n-i} \tag{4.9}$$

$$= \sigma^2_n + \frac{1}{M^2} \sum_{i=1}^{M} \sigma^2_{n-i} \tag{4.10}$$

$$\simeq \sigma^2_n + \frac{1}{M}\sigma^2_n \tag{4.11}$$

For $M = 1$, we would get the same increase in noise as in Equation (4.7) for the plain difference of images. For background maps that average over more than one past frame, however, the noise introduced becomes smaller. Due to the quadratic addition, the increase in noise to the variance $\sigma_{I'_n}$ is less than 5% for $M = 10$. The value of M should not be chosen too large to ensure that the responsiveness of the background map matches the rate of change of the background in the recording.

While the drop in noise compared to simple subtraction of the previous frame is beneficial, a hardware implementation of a sliding average would multiply the number of memory accesses by the number of past images M. These images have to be stored and read every time a new frame arrives. Exponential smoothing is more efficient and has a similar effect.

For exponential smoothing, only one image, namely the previous background map B_{n-1}, has to be kept in the memory of the pipeline. The algorithm is shown in Equation (4.13). To calculate the current background map B_n, only the current image frame I_n and the previous background B_{n-1} are needed. The smoothing is parameterized by the smoothing factor $1/N$.

$$I'_n = I_n - B_n \tag{4.12}$$

$$B_n = B_{n-1} + \frac{1}{N}(I_n - B_{n-1}) \tag{4.13}$$

$$= \frac{1}{N}\sum_{i=0}^{n-1}\left(\frac{N-1}{N}\right)^i I_{n-i} + \left(\frac{N-1}{N}\right)^n B_0 \tag{4.14}$$

The sequence B_n is geometric; its explicit definition is given in Equation (4.14). We use it to calculate the increase in noise for I'_n after background subtraction (Equation 4.18). The influence of each frame on B_n decreases over time as new frames arrive, and once again, we assumed that the background and its width σ_{n-i} do not vary too much within the last N frames. When compared to the sliding average, exponential smoothing has a similar effect on the noise after background subtraction for $N = 2M$.

$$\sigma_{I'_n}^2 = \sum_{i=0}^{n-1}\left(\frac{\partial I'_n}{\partial I_{n-i}}\right)^2 \sigma_{n-i}^2 + \left(\frac{\partial I'_n}{\partial B_0}\right)^2 \sigma_{B_0}^2 \tag{4.15}$$

$$= \sigma_n^2 + \frac{1}{N^2}\sum_{i=0}^{n-1}\left(\frac{N-1}{N}\right)^{2i}\sigma_{n-i-1}^2 + \underbrace{\left(\frac{N-1}{N}\right)^{2n}\sigma_{B_0}}_{\rightarrow 0 \text{ for } n \rightarrow \infty} \tag{4.16}$$

$$\simeq \sigma_n^2 + \frac{\sigma_n^2}{N^2} \frac{1 - \left(\frac{N-1}{N}\right)^{2n}}{1 - \left(\frac{N-1}{N}\right)^2} \tag{4.17}$$

$$\simeq \sigma_n^2 + \frac{2}{N}\sigma_n^2 \tag{4.18}$$

The proposed background treatment can be seen as a high-pass filter applied to each pixel over the time of the recording. Filtering frequencies within each individual frame over *space* would be more difficult: The Airy disk profile of a spot closely resembles a Gaussian distribution, which is mapped to a Gaussian distribution in the frequency domain again. Its most defining frequencies are close to zero, and overlap with the spectrum found in the background map. Removing the background for each frame therefore would impact the signal of the spot as well.

As a last optimization, the calculated background map can be limited to not rise too fast. When a spot is seen of intensity I_{spot}, it would otherwise increase the background map at its position by I_{spot}/N. An image is only expected to vary within about $\sigma_{n,\text{estimated}} = \sqrt{B_{n-1}}$, and the following rule takes this into account:

$$B_n = B_{n-1} + \frac{1}{N} \min\left((I_n - B_{n-1}), \sqrt{B_{n-1}}\right) \tag{4.19}$$

The entire process can be seen in Figure 4.6. Note that all calculations rely on the fact that the value of each intensity is close to the number of photons that hit the CCD sensor at this pixel. For cameras that do not follow this specification, the intensity must be adjusted in advance to make this assumption true.

4.1.5 Spot Detection

After the background has been subtracted, it has become much easier to identify the spots in the image. A spot must have an intensity above a certain threshold, and its center is a local maximum. For the threshold, we can use the estimated variance of the background noise $\sigma_{n,\text{estimated}} = \sqrt{B_{n-1}}$, again for each pixel, and define a multiple of it as the lower limit the center of a spot has to reach. For a threshold of $4\sigma_{n,\text{estimated}}$, less than 0.004% of all pixel intensities is expected to reach the boundary, resulting in about two false positives per 256×256 px frame. The false positives are expected to be removed in the feature extraction step because they do not resemble an Airy disk.

For the determination of the local maximum, a comparison of every pixel with its neighboring pixel is sufficient. This can be done efficiently on a DFE with a line buffer for the previous and the next line of the current

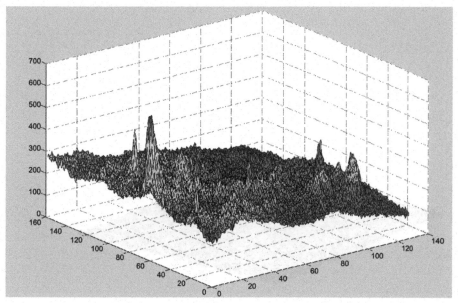

(a) recorded input frame I_N

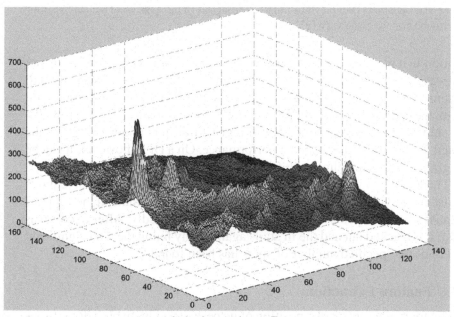

(b) background map B_n

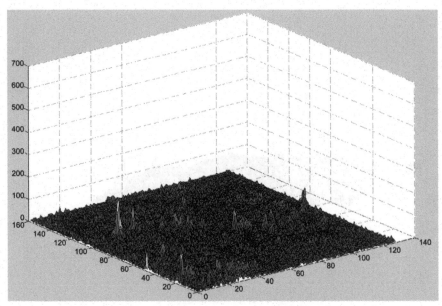

(c) frame I'_n with spots after saturated subtraction

Figure 4.6   Inhomogeneous background handling. The background in an individual input frame is too high and inhomogenous for spot detection. The background map is created with exponential smoothing over previous frames. The signals of the spots stand out only after the background map was subtracted [92].

position in a pipeline, pixel by pixel. However, the true center of spot is not necessarily at the same pixel where the intensity peaks due to noise that is still present in the image. To provide good initial parameters to the following fit, the image can be blurred for this purpose only with a small stencil. On a CPU, this step would increase the number of register transfers proportionally to the stencil size and would impact the performance. On a DFE, however, it can be integrated into the pipeline with few additional resources.

The MATLAB version does not use a threshold to detect spots, but requires the user to provide the total number of spots expected per frame. It then applies a Gaussian least-square fit on the entire frame repeatedly. After a spot was fitted, it is blacked out and the fit starts again, likely fining the next significant spot until the specified number of spots has been found.

4.1.6 Feature Extraction

The features of a spot are the position of its peak, its width in both dimensions and the total intensity. Additionally, the confidence of the position must be known. All parameters can be obtained by fitting, i.e. adjusting the parameters of a 2D Gaussian profile as close as possible to the data. Two popular measures exist to estimate how close a function can be fitted to the data. These are used by the least-square method and the maximum likelihood method, respectively.

Least square: For the least-square method, the differences between fit function and data are squared and added, resulting in a scalar value that is then to be minimized [106]. The difference between a data point and the fit function is also known as the residual. In our application, a data point is given by the intensity $q_{xyy} = I_n(x, y)$ of a pixel at position (x, y) in a spot, and the fit function is the Gaussian profile. The scalar to minimize is χ^2 with respect to the center $\mu_{x,y}$, width $\sigma_{x,y}$ (not to be confused with the variance of the noise), the total intensity Q of the spot, and, if needed, a background scalar B.

$$\vec{a} = (Q, \mu_x, \mu_y, \sigma_x, \sigma_y, B) \tag{4.20}$$

$$\chi^2 = \sum_{(x_i, y_j) \in \text{ROI}} (q_{i,j} - f_{\vec{a}}(x_i, y_j))^2 \tag{4.21}$$

$$f_{\vec{a}}(x, y) = \frac{Q}{2\pi\sigma_x\sigma_y} e^{-\frac{1}{2}\left(\left(\frac{x-\mu_x}{\sigma_x}\right)^2 + \left(\frac{y-\mu_y}{\sigma_y}\right)^2\right)} + B \tag{4.22}$$

χ^2 cannot be minimized analytically. Instead, the minimum has to be found numerically. The optimal set of fit parameters $\vec{a}$ is approached by either following the steepest descent of χ^2 or by approximating χ^2 as a quadratic function and jumping to its assumed minimum [107]. In the first case, the parameter set $\vec{a}$ is modified by $\delta\vec{a}$ along the direction of the slope of χ^2.

$$a'_l = a_l + \delta a_l \tag{4.23}$$

$$\delta a_l = \text{constant} \times \left(-\frac{1}{2}\frac{\partial\chi^2}{\partial a_l}\right) \tag{4.24}$$

$$= \text{constant} \times \sum_{(x_i, y_j) \in \text{ROI}} (q_{i,j} - f_{\vec{a}}(x_i, y_j))\frac{\partial f_{\vec{a}}(x_i, y_j)}{\partial a_l} \tag{4.25}$$

In the second case, we develop χ^2 to the second order and use the Hessian matrix D to jump from $\vec{a}$ to the (assumed) minimum at $\vec{a}'$ directly.

$$\chi^2(\vec{a}) = \chi^2(\vec{a}_0) - \frac{\partial\chi^2}{\partial\vec{a}} \cdot \vec{a} + \frac{1}{2}\vec{a} \cdot D \cdot \vec{a} + \ldots \qquad D_{kl} = \frac{\partial^2\chi^2}{\partial a_k\partial a_l} \tag{4.26}$$

$$a'_l = a_l - \sum_k D_{kl}^{-1} \frac{\partial \chi^2(\vec{a})}{\partial a_k} \tag{4.27}$$

Both methods have different realms where they perform best. Far away from the minimum, the function is unlikely to still be approximated well with a quadratic equation, and the steepest descend method performs better. When close to the minimum, χ^2 can be expected to be smooth enough and the second method is expected to converge faster.

The Levenberg-Marquardt method [106] combines both approaches continuously and has become a standard method for least-square fits. For this, the Hessian matrix D in Equation (4.27) is substituted by $A = D + \lambda 1$. This way, the "constant" in the steepest descend method is removed. For λ approaching zero, the Levenberg-Marquardt method becomes the same as the quadratic approximation. For $\lambda \gg 1$, the diagonal elements of A dominate and the method follows the steepest descent. λ is modified at runtime by the fit algorithm to grow when χ^2 increases, and to shrink when χ^2 decreases until χ^2 almost stops decreasing. Then, the fit has reached an optimum. The confidence of the individual fit parameters can be taken from the diagonal elements of D^{-1}, which acts as the covariance matrix.

MATLAB supports the Levenberg-Marquardt method with the `lsqcurvefit` library function [108], and it is used to fit spots in the MATLAB version that was used as a starting point for accelerating the analysis of the raw data from localization microscopy. The fit function includes the constant background offset B as well as two extra background parameters to fit a constant slope in the x and y direction. This algorithm for fits is compute intense: To evaluate χ^2 or one of its derivatives, the exponential of the fit function has to be calculated for every pixel in the ROI. Additionally, the inverse of the Hessian matrix has to be obtained for the quadratic approximation, which is an operation of at least the order of $o(n^2 \log n)$ for small matrices with n being the number of fit parameters. Last, but not least the total number of iterations $\vec{a} \rightarrow \vec{a}'$ before the stopping condition is reached is not known in advance.

Maximum likelihood: The maximum-likelihood method follows a different approach than minimizing χ^2. Instead, it introduces the *likelihood L* of a distribution to fit the data points [109]. It is defined for two dimensions in Equation (4.29) and maximized by adjusting the fit parameters $\vec{a}$ for the fit. By calculating the likelihood, we can determine how probable the hit of a pixel at position $\vec{x}_{i,j} = (x_i, y_i)$ in the camera by a photon y_i is to happen. For multiple photons, these probabilities are multiplied:

$$L = \prod_{\gamma_i} f_{\vec{a}}(x_i, y_i) \tag{4.28}$$

$$l = \ln(L) = \sum_{\gamma_i} \ln(f_{\vec{a}}(x_i, y_i)) \tag{4.29}$$

Maximizing L requires us to derive the product with respect to the fit parameters. Since sums are easier to differentiate than products, and the exponential in the fit function is removed by the natural logarithm, the function l is preferred over L for maximization. The logarithm is a strictly increasing function and therefore does not affect the solution of the minimization.

We rely on the background removal in the previous step, set $B = 0$ in the fit function, and continue with a Gaussian distribution only scaled by the total intensity. We determine the maximum of l by deriving the fit function with respect to the parameter set and obtain the following system of equations for $\vec{\mu}, \sigma_x$, and σ_y [109]. Note that γ is the index for photons, while i, j labels the position $\vec{x}_{i,j}$ of the pixel that was hit. Equations (4.23) and (4.34) do not give the total intensity Q. It cannot be derived from the maximization, but is already known to be the sum of all photons. The equations for the center $\vec{\mu}$ and the width of the spot (σ_x, σ_y) are known as the Gaussian estimator, as the center of mass or the centroid. They provide an analytic, easy to compute solution to the problem of performing a maximum likelihood fit with a Gaussian profile, given that the background does not have to be fitted as well.

$$\frac{\partial l}{\partial \vec{a}} \overset{!}{=} \vec{0} \qquad\qquad Q = \sum_{\gamma} 1 = \sum_{x,y} q_{x,y} \tag{4.30}$$

$$\frac{\partial l}{\partial \mu_x} = \sum_{\gamma} \frac{x_\gamma - \mu_y}{\sigma_x} = \sum_{i,j} q_{i,j} \frac{x_i - \mu_x}{\sigma_x} \overset{!}{=} 0 \tag{4.31}$$

$$\Rightarrow \mu_x = \sum_{i,j} \frac{q_{i,j}}{Q} x_i \qquad\qquad \mu_y = \sum_{i,j} \frac{q_{i,j}}{Q} y_j \tag{4.32}$$

$$\frac{\partial l}{\partial \sigma_x} = -\frac{Q}{\sigma_x} + \sum_{\gamma} \frac{(x_\gamma - \mu_x)^2}{\sigma_x^3} \overset{!}{=} 0 \tag{4.33}$$

$$\Rightarrow \sigma_x^2 = \sum_{i,j} \frac{q_{i,j}}{Q} (x_i - \mu_x)^2 \qquad \sigma_y^2 = \sum_{i,j} \frac{q_{i,j}}{Q} (y_j - \mu_y)^2 \tag{4.34}$$

The error of the localization $(\Delta\mu_x, \Delta\mu_y)$ is obtained through error propagation (Equation 4.35). For the pixelation error, we find $\Delta x^2 = \Delta y^2 = 1/12$ when we integrate a constant probability function over the width of a pixel. The variance of $q_{x,y}$ is $\Delta q_{x,y} = \sqrt{q_{x,y} + B_{x,y}}$ for the assumed Poisson noise. As expected, the localization error $(\Delta\mu_x, \Delta\mu_y)$ decreases if the total number of photons Q rises and increases for high background levels $B_{x,y}$.

$$\Delta\mu_x^2 = \sum_\gamma \left(\frac{\partial\mu_x}{\partial x_\gamma}\right)^2 \Delta x_\gamma^2 + \sum_{i,j} \left(\frac{\partial\mu_x}{\partial q_{i,j}}\right)^2 \Delta q_{i,j}^2 \qquad (4.35)$$

$$= \frac{1}{12Q} + \sum_{i,j} \left(\frac{x_i - \mu_x}{Q}\right)^2 (q_{i,j} + B_{i,j}) \qquad (4.36)$$

$$\Delta\mu_y^2 = \underbrace{\frac{1}{12Q}}_{\text{pixelation}} + \underbrace{\sum_{i,j} \left(\frac{y_j - \mu_y}{Q}\right)^2 (q_{i,j} + B_{i,j})}_{\text{Poisson noise}} \qquad (4.37)$$

When fitting a Gaussian, we expect the pixels remote to the center $\vec{\mu}$ to have vanishing values $q_{x,y}$. If not, we can already see that the equation for the center $\vec{\mu}$ of the Airy disk is susceptible in this case: The term $(q_{x,y}/Q)x$ wrongfully draws the calculated center toward the nonvanishing pixel. In localization microscopy, this might happen at two occasions, either due to noise in the image or by the tail of a second spots that leaks into the neighborhood of the spot to be fitted.

To address noise, we may simply subtract, e.g., two times the width of the noise, set any negative values to zero, and thereby remove 99% of the noise where the signal should be zero. The effect of removing the background and additionally performing a saturated subtraction of $2\sigma_{\text{noise}}$ can be seen in Figure 4.7. Setting values to zero near the border is equal to excluding pixels from the calculation of the location and the width, and to tailoring the ROI to the shape of the spot.

For the effect of other spots moving the center toward where their tail touches the current spot, we can find the minimum in intensity between both and set all values behind it to zero that belong to the other spot. Setting values of $q_{i,j}$ to zero removes them from the sum in Equation (4.32), excluding them from the fit.

The size of the quadratic ROI where the fit takes place in must be chosen between two extremes. A ROI too small will cut off valuable information; it

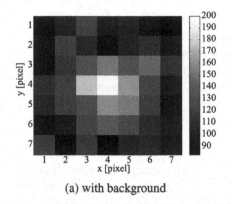

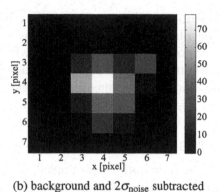

(a) with background

(b) background and $2\sigma_{\text{noise}}$ subtracted

Figure 4.7 A simulated noisy spot with total charge $Q = 1000$, background level $N_B = 100$, and width $\sigma_{x,y} = 1.4$ px at $\vec{\mu} = (4.1, 4.2)$. By setting pixels with $q_i < 2\sigma_{\text{noise}}$ to zero, the feature extraction becomes robust against ROIs with nonvanishing values at the border. Here, the center was determined with a subpixel accuracy of 0.16 px or 16 nm [92].

must therefore be at least twice the width of the spot. A ROI too big, however, may include noise at the borders and increase the chance that a close-by spots leaks into the ROI, disturbing the fit. We will examine the fine-tuning of the entire algorithm later to find the optimal parameters for localization microscopy with visible light.

4.1.7 Super-Resolution Image Generation

Finally, when all spot locations have been computed from the recording, the data are visualized by plotting the spots into a new image with a higher resolution. The pixel size of an optical camera is build close to the optical resolution, as a finer resolution would decrease photon efficiency and increase noise. For a recording with a resolution of 256×256 pixels, it is therefore common to increase the resolution of the final plot by a factor of ten, much like the precision improvement that is an order of magnitude. A super-resolution image will then have a resolution of 2560×2560 pixels.

In the most simplest case, the positions are plotted as single white dots in the final image. A better plotting technique that was used in the localization image shown in Figure 4.1b is to blur each dot with its confidence before. This way, the localization error is visible to the human eye. Other imaging techniques also require the distance of its nearest neighbor(s) to determine the blur radius for visualization.

With recordings typically having tens of thousands of fluorophore position to be plotted, the visualization may take a considerable amount of time. Each position must be blurred, which is implemented using a Gaussian distribution to draw the gray values around a spot position. For this, the Gaussian function must be calculated repeatedly. The total time for image generation can approach multiple minutes.

4.1.8 State of the Art

Starting from initial algorithms that relied on the least-square fit for localization, a couple of solutions have been developed to mitigate the long hours consumed by the analysis. The approaches for acceleration so far can be split into improvements of the algorithms and portings to special hardware.

The center-of-mass calculation as presented before was examined before with regard to the selection of a suitable threshold and the exclusion of noisy pixel values for localization [110]. The background must be determined and subtracted manually before analysis can start. A flavor of the presented center-of-mass algorithm is also available as a plug-in for the free software ImageJ [111].

The least-square fit and the Gaussian estimator, although seemingly different algorithms, can be combined using the previously known Gaussian profile as a mask of weights $M_{x,y}$ for the calculation of the center of mass [112]. The center of the mask must be the same as the center of the spot $\vec{\mu}$, and the algorithms use an iterative approach until both positions are similar (Equation 4.38). As an advantage, noisy pixel values at the boundary are continuously masked out from the calculation of $\vec{\mu}$. Still, the algorithm is iterative and therefore time-consuming for a large number of spots.

$$\mu_x = \frac{\sum\limits_{i,j} x_i q_{i,j} M_{i,j}}{\sum\limits_{i,j} q_{i,j} M_{i,j}} \qquad (4.38)$$

The fluoroBancroft algorithm was developed as a noniterative alternative to least-square fitting [113]. Its accuracy is comparable to a Gaussian fit for spots covering only a small number of pixels with a low background level and a high signal-to-noise ratio. The width of the spot must be known in advance, but can usually be measured or computed before, since all spots from the focus area should share the same profile only scaled by the total intensity. The development of the algorithm was inspired by the also closed-form (noniterative) Bancroft's algorithm used for position determination in the Global Positioning System (GPS) [114]. The calculation of a localization with the fluoroBancroft algorithm requires the natural logarithm for each pixel and a matrix inversion to be evaluated. Equation 4.39 gives the required algorithm for the computation of a localization $\vec{\mu}$ in two dimensions with intensity I_i and background $N_{i,Bg}$ at position (x_i, y_i).

$$\vec{\mu} = \begin{pmatrix} 1 & 0 & 0 \\ 0 & 1 & 0 \end{pmatrix} ((B^\mathrm{T} B)^{-1} B^\mathrm{T}) \vec{\alpha} \tag{4.39}$$

$$\alpha_i = \frac{1}{2}(x_i^2 + y_i^2 + 2\sigma^2 \ln(I_i - 2N_{i,\mathrm{Bg}})) \qquad B = \begin{pmatrix} x_1 & y_1 & 1 \\ & \vdots & \\ x_n & y_n & 1 \end{pmatrix} \tag{4.40}$$

It is beyond the scope of this book to provide the derivation of the equations, but we can see that the logarithm cannot be evaluated for pixel values with a background level of half the intensity or above. These values must at least be removed from the set of pixels. The algorithm does not provide the confidence of the extracted localization, but it was shown that it scales with the inverse square root of the total number of photons [115], similar to the Gaussian estimator.

As a summary of the algorithms, all methods provide very similar results for well-defined signals from spots with low noise and a high total intensity. For weak signals, center-of-mass and least-square fits introduce a large bias, but behave differently [116]. Least-square fits tend to not converge or to produce a fit with a center far off the true location. On the other hand, center-of-mass algorithms are biased toward the center of the ROI that was used to cut out the spot. Center of mass also shows a bias toward the center if the background has not been subtracted completely, leading to pixel locking in the final image; that is, every location is biased toward the center of the underlying pixel. It is therefore of high importance to estimate the background level correctly. Least-square fits can compensate for homogeneous backgrounds with additional fit parameters and are therefore easier to deploy at the cost of computing time.

The runtime of localization algorithms so far has only been sped up with graphics cards as application accelerators. The *MaLiang* method [117] is a port of an iterative maximum-likelihood fit. Since the implementation includes the background level in the fit as a parameter, it does not resolve to the Gaussian estimator. Despite its iterative nature, it provides real-time capability even for fast cameras. Compared to fluoroBancroft, the analysis of a typical frame is faster by more than a factor of ten on an NVIDIA GeForce 9800GT graphics card, and faster by more than a factor of 20 than a Gaussian least-square fit in software.

4.1.9 Analysis of the Algorithm

The analysis of the algorithm for localization microscopy started with an implementation in MATLAB that was provided by the research groups of Prof. Christoph Cremer from the Institute of Molecular Biology Mainz and Prof. Michael Hausmann from the Kirchhoff Institute for Physics at Heidelberg University. It is embedded in a suite of tools for analysis and visualization.

The algorithm consists of a chain of operations. First, the recording is read, encoded as a stack of frames in either the Khoros standard data format (KDF) or the tagged image file format (TIFF) format. For imagery with a high inhomogeneous background level, an option is provided to perform the analysis on the difference of the current with the previous frame, increasing the noise level as described in Section 4.1.2 above. An iterative least-square fit with a Gaussian profile is used for the feature extraction. Finally, the super-resolution image is generated by plotting the locations and blurring them with their fit error.

As a starting point for acceleration, the program was profiled with the MATLAB profiling tool. Since profiling slows down the execution, an image stack with only few frames was used. The result can be seen in Figure 4.8. The operation that consumes the most computing time is the fit function `nonlin2` that performs a least-square fit on the ROI of each spot. It computes the Gaussian function 26 times on average per spot, which is returned by `gauss2dbPL3D` for a given position. Afterward, the fit has converged or was aborted. Both functions consume about 56% of the total compute time.

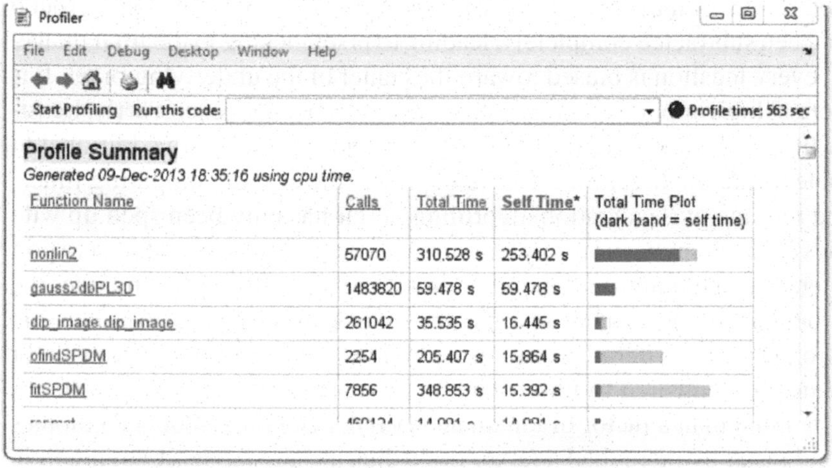
Profiler
File Edit Debug Desktop Window Help
Start Profiling Run this code:
Profile time: 563 sec
Profile Summary
Generated 09-Dec-2013 18:35:16 using cpu time.
Function Name
Calls
Total Time
Self Time*
Total Time Plot (dark band = self time)
nonlin2 57070 310.528 s 253.402 s
gauss2dbPL3D 1483820 59.478 s 59.478 s
dip_image.dip_image 261042 35.535 s 16.445 s
ofindSPDM 2254 205.407 s 15.864 s
fitSPDM 7856 348.853 s 15.392 s

Figure 4.8 Profiler result of the MATLAB implementation for localization microscopy.

dip_image.dip_image is a converter function that makes a frame from the TIFF image stack available for the DIP library functions, a collection of image processing function to simplify common operations [118]. It copies the data and can be eliminated with our own implementation of a TIFF library. The remaining functions have a small self time: ofindSPDM searches for spots in an image and fitSPDM is the top-level function.

Overall, a big amount of time is spent in the least-square fit that could be saved with the noniterative Gaussian estimator. We started with an examination of the fit function and then re-implemented the remaining parts of the application.

4.1.9.1 Methods

While a DFE application could certainly accelerate the iterative least-square fit, it seemed that the time-consuming feature extraction could be already accelerated by moving this part of the algorithm to a fit with a closed form. Since a modification is very likely to change the properties of the extracted locations, the precision of all known closed-form fits had to be evaluated, notably the Gaussian estimator and the fluoroBancroft algorithm.

The accuracy of a fit is given by its fit error, in our case the difference between the true location of the spot and the extracted center of the Gaussian profile. However, real data coming from the microscope cannot be used for this purpose, as the true locations are unknown. It is therefore necessary to generate artificial spots that follow the characteristics of real spots. In particular, the noise of the background and the spot have to be simulated as closely as possible to the real-world data.

The Monte Carlo method provides such a simulation [119]. It combines deterministic elements such as the true position, the width, and the total intensity of a spot with random elements for noise generation. The noise was known to be dominated by Poisson noise and could be generated accordingly with a pseudo-random number generator with the undisturbed profile of an ideal spot as the input. After a large number of simulated spots and performed fits, the average fit error and its distribution could be determined. For each set of width (σ_x, σ_y), total intensity Q, and background level B, 2000 virtual frames were simulated.

The original algorithm was already present in MATLAB, and therefore, the simulation environment was written in the MATLAB programming language as well. The aim of this process was to find an algorithm that maintains accuracy and at the same time can be implemented efficiently in the dataflow model on an FPGA.

The accuracy of the entire algorithm does not only depend on the fit algorithm alone, but also on the starting parameters and, to an even greater extent, to the ability to identify spots on a noisy background in the first place. Therefore, a sufficiently sized frame had to be simulated. The true position was varied such that it occupied all positions within the boundaries of a pixel with equal probability, as a fit may have a bias toward the center of a pixel or the intersection between pixels (pixel locking).

For the background removal, several approaches were examined. Iterative fits and fluoroBancroft are able to include the background into the fit, while the Gaussian estimator relies on the subtraction of the background in advance.

After simulation, the following algorithm proved suitable in both accuracy and dataflow for implementation on a DFE:

1. **Background elimination:** Exponential smoothing was chosen to determine the background map $B_n(x, y)$ for frame n. The rise of the background was limited to the width of the Poisson noise $\sqrt{B_{n-1}}$ to limit its rise when a spot is encountered. The smoothing factor was chosen to be $N = 8$, large enough for typical spots that are visible only in a single frame before becoming dark again, but small enough to follow changes in the background caused by bleaching.

$$B_n = B_{n-1} + \frac{1}{N} \min\left((I_n - B_{n-1}), \sqrt{B_{n-1}}\right) \qquad (4.41)$$

2. **Spot finding:** After the background was subtracted and any negative values have been set to zero, a copy of the resulting image is made and blurred by convolving it with a stencil $S = 1_{33}/9$, that is a 3×3 matrix where each element equals $1/9$. While the copy is unsuitable for feature extraction after blurring, this step is beneficial for spot finding. The fluctuations of the noise are greatly reduced, and the true center of a spot is more likely to be at the pixel with the highest value. For a spot to be detected, its center pixel must be a local maximum and above a certain threshold, given in multiples of the width of the noise level. This blurring improves the centering of the ROI for the spot, which is then forwarded to the next processing step. For the given combination of microscope and CCD camera, a 7×7 ROI was chosen.

3. **Spot separation:** The ROI may also contain the signal of more than one spot. A signal of the main spot in the center of the ROI may be disturbed by a second spot close by that leaks into the ROI with its tail. If the disturbance is not too large, the signal of the second spot can be removed

well by removing all pixels beyond a local minimum when radially scanning from the center to the borders of the ROI. Spot separation does not improve the signal of the spot, but it helps the following feature extraction. Therefore, the confidence of the fit must be calculated without spot separation.

4. **Zero suppression:** Noise left over in the ROI from the previous step reduces the quality of the fit, especially when it increases the value of pixels at the boundary of the ROI for the center-of-mass calculation in the Gaussian estimator. We therefore suppress all values smaller than two times the width of the Poisson noise by subtracting all values in the ROI with $2\sigma_B = 2\sqrt{B_n}$ and then setting all negative values to zero.

5. **Feature extraction:** The Gaussian estimator is used to determine the properties of the spot. For the center ß, the ROI with zero suppression is used. All other properties are determined with zero suppression disabled.

6. **Super-resolution image generation:** The drawing of the final image is left to the software on the CPU. It can still be accelerated by calculating the profile of a location blurred by its confidence only once for a given width casted to an integer. After the first calculation, its bitmap is cached in a hash map. For further spots with the same confidence, the bitmap can be retrieved again and be drawn on the image with a saturated add. The discretization of the confidence is chosen fine enough to be invisible to the human eye.

Since MATLAB is an interpreted computer language, its runtime is susceptible to the degree of library function versus interpreted code found in the program. The libraries for core functions such as matrix operations are implemented natively and have been heavily optimized, but the gained speed can be easily lost in the scripted part [120]. To ensure a fair comparison between software on CPUs and DFE acceleration with FPGAs, the algorithm presented above was also implemented in the C++ programming language.

4.1.9.2 Dataflow

The algorithm reads the image stack frame by frame. For the stages for background elimination and spot finding, the unit of processing is the frame, accessed pixel by pixel in x and y direction. Background elimination contains a loop in the dataflow, since the background map for the current frame depends on the data of the previous background map (Equation 4.41). Therefore, an entire background map has to be buffered and read back later with a delay equal to the number of pixels in one frame.

For spot finding, the dataflow branches into two copies of the stream. One stream is blurred and used for spot finding as described above, looking for a local maximum above a threshold. The other stream is unmodified and, if a spot has been found, used to create a 7×7 pixel ROI with the spot in its center. Here, the unit of processing changes from entire frames to small ROIs. Because ROIs can overlap for close-by spots, pixel values may need to be duplicated, and its ROIs must be buffered into a queue for further processing.

The following steps from spot separation to feature extraction operate on ROIs, until finally a location is produced with the other features of the spot attached. Zero suppression is a point function and therefore easily integrated. For spot separation, the ROI must be scanned from the inside to the outside to find local minima and set all pixel values behind to zero. This can be done by scanning the ROI horizontally, vertically, and horizontally again. For each direction, every pixel has to be read and potentially set to zero once. The access pattern is shown in Figure 4.9 and differs from the usual sequential pattern from left to right and from top to bottom. However, since the latency is still restricted to a ROI and only the access order within the ROI is changed, it can be embedded into the same dataflow without the need to break the pipeline apart.

The final step on hardware, feature extraction, relies on accumulators to calculate the total intensity Q, the location $\vec{\mu}$, the width (σ_x, σ_y), and the location confidence $(\Delta\mu_x, \Delta\mu_y)$. All accumulators sum over each pixel

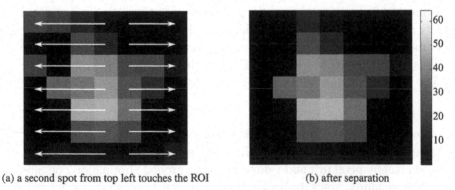

(a) a second spot from top left touches the ROI   (b) after separation

Figure 4.9   The access pattern of the spot separator is from center to left and right and from top to bottom. Every pixel value is read only once. Values that are past a local minimum in are set to zero. The process is then repeated in vertical direction and in horizontal direction again [92].

exactly once, and its commutative property allows an arbitrary order. After each pixel value has been examined, the properties are submitted to the output as a long vector. Overall, the dataflow of the entire algorithm concentrates the data in two steps. It starts with a stack of frames where first ROIs with spots inside are filtered out, and finally, only seven values are produced for each location.

4.1.9.3 Dimensioning of the hardware

The knowledge about the dataflow already gives us a description close to the actual implementation of the hardware. The last bits of missing high-level design are the type of parallelism to use and the data types for the multiple variables.

Parallelism: The operators in the dataflow are applied to two different entities: First entire frames until the spots have been found, and later, ROIs cut out from the frames for feature extraction. This requires us to split the design into two statically scheduled pipelines that run a different number of clock cycles. The first pipeline consumes one pixel of a frame per clock cycle, and the total number of clock cycles is therefore equal to the number of pixels in the stack of frames. The second pipeline operates on 7×7 pixel ROIs, also with one pixel per clock cycles. Since ROIs can overlap, the number of pixels in ROIs could be larger than the total number of pixels for pathological recordings. In practice, however, a single frame contains only few spots and the second pipeline is expected to run only for a fraction of the time of the first one.

In Figure 4.10, a sketch of both pipelines can be seen. In the language of the MaxCompiler library, these are called kernels. The first kernel contains all frame operations and the second kernel extracts the features from each ROI.

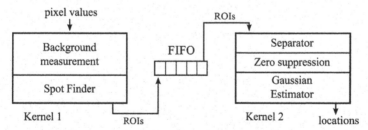

Figure 4.10 The pipeline design. Each pixel from the camera is fed through the first kernel, where the background gets measured and subtracted. When a spot is found, its ROI is sent to the second kernel. Here, the separator removes values from nearby spots that touch the ROI. The estimator finally calculates the position data. Both kernels are decoupled by a FIFO [92].

The FIFO between both kernels stores the pixel values of each ROI per entry, along with its properties such as its x and y position in the frame, the level of the background that has already been deduced and the frame number. The FIFO implements the flexible interconnect to allow both kernels to run with a different number of active clock cycles.

The parallelism found here is pipeline (MISD) parallelism. All operations are performed in parallel, but the pipelines only have a single instance each. The total expected throughput is one consumed pixel value per clock cycles. For an estimated minimal clock frequency of 100 MHz, a single frame with 256×256 pixels would compute within 0.7 ms, which is much faster than the camera can produce them. Typical recording times per frame are between 100 and 10 ms, which is one or two orders of magnitude slower than the proposed hardware design. Multi-pipe (SIMD) parallelism for additional acceleration could be done by instantiating the first pipeline several times and splitting each frame among them, but is beyond the requirements of the application.

Numerics: The intensities are typically recorded with an accuracy of 12 bits, and the cameras used for microscopy in the co-operating research groups of Prof. Christoph Cremer and Prof. Michael Hausmann followed this rule. For further processing, the data are stored in frames with 16 bit per pixel values, making it immediately compatible with general-purpose computing. The input of the system is therefore an unsigned 16-bit integer with only the bottom 12 bits containing information. The chosen number encodings can be seen in Table 4.1. They follow from the input encoding and the steps of the algorithm.

Table 4.1   Numerical encoding used for the localization algorithms on hardware. Except for the last step, all operations can be done with fixed-point encodings, leading to a smaller resource footprint than floating point (see also Figure 3.11)

Use Case	Number Encoding
Input pixel values	Unsigned 16-bit integer encoding
Pixel location	Unsigned 16-bit integer encoding
Background map	Signed fixed-point encoding with 16 integer and 4 fractional bits
ROI pixel values	Signed fixed-point encoding with 16 integer and 4 fractional bits
Accumulators	Signed fixed-point encoding with 20 integer and 4 fractional bits
Division by Q	Floating-point encoding with single precision
Output location and features	Floating-point encoding with single precision

For background elimination, the pixel values are smoothed exponentially over time and divided by the factor N in the process. With $N = 8$, the resulting numbers will have 3 fractional bits. To also allow for $N = 16$, the number of fractional bits was set to 4. Since the background map will have a similar range as the input, leading bits cannot be omitted. The opportunity to stay with fixed-point numbers is expected to keep the resource footprint small for background elimination and spot findings. The square root to estimate the width of the noise cannot be represented with a fixed number of fractional bits exactly for most inputs. The error it may introduce to the locations will need to be measured later. An unsigned 16-bit integer is sufficient for 256×256 pixel frames to address a pixel in the background map. The pixel values passed in a ROI are obtained by subtracting the background are therefore share the same encoding.

Spot separation and zero suppression are operations that remove pixels or subtract pixel values by numbers of the same encoding and do not require a cast. For the accumulators, such as $\sum_{x,y} x q_{x,y}$ for the x location and similar for width and confidence, non-negative values are multiplied and summed up 49 times for a 7×7 ROI. The number of integer bits has therefore to be increased to 20 bits, taking into account the 12 bit of information found for each pixel and 3 bits for the encoding of x. To obtain the final location, the accumulated values have to be divided by the total intensity Q. For this, a floating-point encoding with single precision was chosen. A location requires 8 bits for the pixel location and at most 10 bits for the subpixel position; the 23-bit fractional part for the significand is oversized by at least 5 bits. However, the standardized format is also used for the output of the pipeline and simplifies communication with the host.

4.1.10 Implementation

The algorithm presented in the previous section was implemented on a DFE in a MaxWorkstation built by Maxeler Technologies. The workstation houses the CPU motherboard with the DFE card on top, connected via PCIe 8x. The CPU embedded in the MaxWorkstation is an Intel Core i5-750 processor with four physical cores at 2.66 GHz and 4 GB of DDR3 RAM.

For the description of the dataflow graph and the generation of the interediate VHDL representation, MaxCompiler 2011.3 was used. The bitfile containing the configuration for the FPGA was synthesized with the Xilinx Integrated Software Environment (ISE) 13.3 for a Xilinx Virtex-5 LX330T FPGA. On the DFE board, 12 GB of RAM is available for access from the FPGA, but was not used for this application.

4.1.10.1 Host code

The application consists of a software and a hardware part. The software controls the hardware and acts as an interface for the data streams. It provides the imagery to the DFE and stores the location it receives on disk. The hardware part is encoded in a bitfile that defines the configuration of the FPGA. It is pushed to the DFE by the software during initialization.

The software program, also known as host code because it runs on the CPU of the host system, was written in C++ on Linux and executes the parts of the algorithm that can hardly be accelerated with a DFE. After it configured the FPGA with the bitfile, the host code sets the parameters of the algorithm such as the total number of images, the threshold for the spot finder, and the total number of cycles the pipeline should run. For localization microscopy, it converts the image data from TIFF or KDF-encoded recordings into a three-dimensional array of 16-bit pixel values and streams them continuously to the DFE. On the receiving side, the locations are visualized in a super-resolution image or can be stored in a comma-separated values (CSV) file for further processing.

In a pipelined design, data are sent to and received from the hardware simultaneously. The host code therefore has to perform two I/O operations at the same time. After the hardware was initialized, two POSIX threads are created, one for each direction of the dataflow. The data are sent from and received to a buffer in the host memory through calls to a C library offered by Maxeler Technologies. Internally, the contents of the buffers are sent to and received from the hardware through DMA transfers on the PCIe bus.

Before the implementation of the hardware was started, the functionality was tested with a MATLAB program that closely resembles the hardware design for bugs and accuracy of the calculation. The hardware part consists of the compute-intense tasks in two statically scheduled pipelines as shown in Figure 4.10, and its implementation will be described below.

4.1.10.2 Background removal

To remove the background, a map of its level has to be estimated and subtracted from each frame. The pipeline for this process can be seen in Figure 4.11. The current background map is stored in BRAM and hold exactly one frame. Hence, the BRAM acts as a ring buffer for the background map.

The read and write addresses are generated by counters that are incremented with each clock cycle and wrap around if they hit the number of pixels per frame. The part of the pipeline that calculates the background B_n forms a loop with the BRAM. Its latency equals the offset between the address

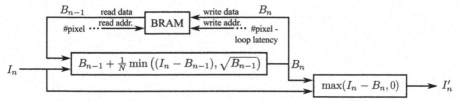

Figure 4.11 Hardware implementation of the background removal. The design of the statically scheduled pipeline consists of a BRAM buffer that holds the background map, and the pipelined implementation of the formulas that estimate and subtract the background pixel by pixel.

generators for read and write access. With the square root being the most compute-intense operation, the latency of the loop is 48 clock cycles.

4.1.10.3 Spot detection
After the background has been subtracted, a blurred copy of the resulting image is made and every pixel is tested whether it is a local maximum and whether its value exceeds the threshold. The source code for the detection of a maximum is shown in Listing 4.1. It builds a pipeline that accepts one pixel value per clock cycle and produces a stream of one-bit Boolean values that indicate whether a maximum was found at the current pixel position.

To determine whether a value is maximal in its environment, a 7×7 pixel ROI is created from the input stream in the first set of nested for-loops. Negative stream offsets are allowed by MaxCompiler and are resolved when the design is globally scheduled at the beginning of a build. In the second set of nested for-loops, every value in the ROI is compared to the central value. If all other values are less or equal than the central value, the function returns true. Note the usage of an intermediate result `intermResult` for the comparison between the loops that decrease the latency of the pipeline. It forces the comparator part of the pipeline to form a tree with an intermediate level instead of building an unbalanced tree with only one long branch. A similar approach was chosen for the blurring with a 3×3 pixel matrix.

```
KArrayType<DFEVar> roiType = new KArrayType<DFEVar>
    (pixelFracValueType, roiSize);

DFEVar isLocalMax(DFEVar pixel) {
    // Build a ROI around the current pixel
    KArray<DFEVar>roi = roiType.newInstance(this);
    for(int y = 0; y < roiEdgeLength; y++) {
```

```
for (int x = 0; x < roiEdgeLength; x++) {
   roi.connect(y * roiEdgeLength + x,
         stream.offset(pixel, x - roiRadius
               + (y - roiRadius) * imgWidth));
}


}
// Does "pixel" have the maximum value in the ROI?
DFEVar localMax = constant.var(true);
   for(int y = roiRadius - 1; y <= roiRadius +1;
      y++) {
   DFEVar intermResult = constant.var(true);

   for(int x = roiRadius - 1; x <= roiRadius
      +1; x++) {
      int roiIndex = y * roiEdgeLength + x;
      if(y != roiRadius || x != roiRadius) {
      intermResult = intermResult &
            (roi[roiIndex]<= pixel);

   }
}
   localMax = localMax & intermResult;
}
   return localMax;


}
```

Listing 4.1 Function in MaxJ that builds a pipeline for local maximum detection.

When a spot is found, its ROI is sent to the output as a whole, where it will get enqueued into the FIFO for processing by the second kernel. For a 7×7 pixel ROI and 20 bits per pixel, the width of the word is 980 bits wide. Additionally, the image number, the position of the ROI, and the background level are submitted.

When the image stack has been analyzed, the last line of pixels in the last frame is used to generate empty ROIs with the frame number set to a negative number. These ROIs are needed due to limitations of the MaxCompiler software libraries that require the amount of data for a DMA transfer to be known in advance. The empty ROIs pad the last transfer and ensure that the last DMA buffer that still contains some valid locations is flushed and does not cause the final transfer to the host to stall. The invalid locations are identified in the host code by the negative frame number and are discarded.

4.1.10.4 Spot separation

For the separation of close-by spots, the hardware needs to scan the ROI from the center to the outside horizontally, vertically, and horizontally again. Pixel values behind a local minimum are likely to belong to a nearby spot and need to be discarded. Since the Airy disk is flat at the center, noise may trigger such a decision by accident. The inner 3×3 pixels are therefore protected and left unaltered in any case.

The order of access for the horizontal scan is shown in Figure 4.12a. Given a position (x, y) and a ROI with radius r_{ROI} and width $w_{ROI} = 2r_{ROI} + 1$, it can be precomputed before hardware synthesis in MaxJ with the following formula.

$$\text{access order} = y \times w_{ROI} + \begin{cases} r_{ROI} - x, & \text{if } x \leq r_{ROI} \\ x & \text{otherwise} \end{cases} \qquad (4.42)$$

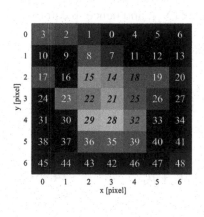

(a) horizontal ROI access order

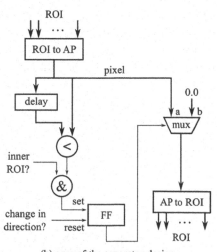

(b) core of the separator design

Figure 4.12 ROI hardware separator. Signals from close-by ROIs are removed by scanning theROI from the center to the outside and setting all values to zero after a local minimum. The values of the inner 3×3 pixels are protected against modification.

The design for one direction is shown in Figure 4.12. First, the pixels in the ROI have to be reordered in the direction of the access pattern (AP). This operation is done by selecting the pixels of the buffered ROI in the order given in Equation (4.42) with a multiplexer (ROI to AP). The current pixel value is then compared with the previous one that has been delayed to arrive in time. If the current pixel value is greater than the previous one, a local minimum has been found. If the pixel is also not an inner one, a flip-flop is set that causes all following pixel values to be set to zero until the direction of the scan is changed. Last, the ROI is buffered again for the following logic.

For the reordering, a counter is needed that tracks the number of the pixel in the ROI. For the decision whether the direction changed or whether the pixel is an inner pixel, a second counter tracks the x position. For the vertical scan, the hardware was adjusted accordingly.

4.1.10.5 Feature extraction
A set of accumulators is used when scanning the ROI pixel by pixel for the determination of the total intensity Q, the location $\vec{\mu}$, the width (σ_x, σ_y), and the location confidence $(\Delta\mu_x, \Delta\mu_y)$. For the width, the formulas were rewritten to not depend on the location for each iteration, such that the accumulators for width and location can work in parallel. The final form for σ_x (Equation 4.46) does not contain $\vec{\mu}$ in the scope of the accumulator $(\sum)$, and hence, ß can be calculated independently and is subtracted only at the end. The equations for μ_y and σ_y are obtained by substituting all x for y.

$$Q = \sum_{i,j} q_{i,j} \qquad \mu_x = \frac{1}{Q} \sum_{i,j} q_{i,j} x_i \tag{4.43}$$

$$\sigma_x^2 = \sum_{i,j} \frac{q_{i,j}}{Q} (x_i - \mu_x)^2 \tag{4.44}$$

$$= \frac{1}{Q} \sum_{i,j} x_i^2 - \frac{2\mu_x}{Q} \sum_{i,j} q_{i,j} x_i + \mu_x^2 \tag{4.45}$$

$$= \frac{1}{Q} \sum_{i,j} q_{i,j} x_i^2 + \mu_x^2 \tag{4.46}$$

The same process was done for the calculation of the confidence. Because zero suppression effectively shrinks the ROI, only summands for pixels with a nonzero intensity after background subtraction are added up.

$$\Delta\mu_x^2 = \frac{1}{12Q} + \sum_{i,j}\left(\frac{x_i - \mu_x}{Q}\right)^2 (q_{i,j} + B) \tag{4.47}$$

$$= \frac{1}{12Q} + \frac{\sigma_x^2}{Q} + \frac{B}{Q^2}\sum_{i,j}(x_i - \mu_x)^2 \tag{4.48}$$

$$= \frac{1}{12Q} + \frac{\sigma_x^2}{Q} + \frac{B}{Q^2}\left(\sum_{i,j} x_i^2 - 2\mu_x\sum_{i,j} x_i + \mu_x^2\sum_{i,j} 1\right) \tag{4.49}$$

All features except for the confidence are calculated on intensities before spot separation. Hence, two sets of accumulators are instantiated: one set for $(\Delta\mu_x, \Delta\mu_y)$, and one set for the remaining features.

The outputs of the accumulators are connected according to the formulas above and operate in units of pixels. After all pixels of a ROI have been calculated, the root of the squared widths and confidences is taken and the result is sent to the host computer through its PCIe interface.

The results of the feature extraction are checked for plausibility in the host code. Otherwise, photon noise that was confused with a spot by the spot finder could end up as a wrong result. The limits on the features for valid spots require the spot to have a total charge $Q > 0$ and a width $\sigma_x, \sigma_y > 0$. The localization accuracy must be better than 0.7 pixels in both directions to ensure a fit with subpixel precision. Finally, the separator was allowed to only remove 30% of the total charge, as our simulations showed this to be the limit for spot separation with sufficient accuracy [92].

4.1.10.6 Visualization

The command line version of the host code produces a list of locations, encoded as CSV, for further analysis. To allow the operator an immediate visual feedback, a graphical user interface (GUI) was implemented. The previous host code was packed into a software library to be called from the GUI. The GUI itself was written with the platform-independent Qt toolkit [121]. A screenshot is presented in Figure 4.13.

With the super-resolution image having a resolution increased by a factor of ten in both directions, the image generation took more time at first than the accelerated analysis. Each location had to be drawn as a Gaussian distribution with the fit confidence as the width. This imposed a high load on the floating-point unit of the CPU. To mitigate the bottleneck, the Gaussian profiles were computed only once for each fit confidence, stored in a cache, and drawn on top of the plot if the same confidence occurred again. The confidence was rounded to an integer in the coordinates of the super-resolution plot; the introduced

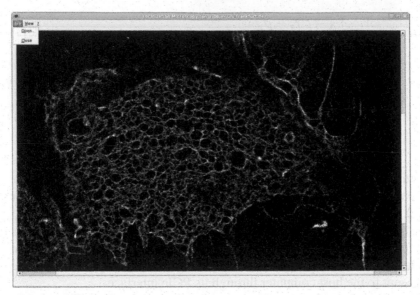

Figure 4.13   Screenshot of the Qt GUI. The program calls the host code as a library function to configure the DFE and to stream the image stack. Afterward, it renders a super-resolution image from the location results.

aberration by rounding remained invisible to the human eye. For the cache, a C++ map (`std::map`) was used to look up the Gaussian distribution with the confidence as the key.

4.1.11 Results

For localization microscopy, the most important benchmarks are the accuracy of the localization and the number of spots found. Both influence the resolution of the final localization image directly. A high fit accuracy is needed to improve the precision of the Abbe diffraction limit, and a high number of locations are necessary to allow the human eye to recognize the structures as seen with a conventional light microscope. While an improvement in speed was the primary motivation for this book, we need to ensure first that the accuracy of the results was preserved.

4.1.11.1 Accuracy

The fit errors presented here were obtained by performing a Monte Carlo simulation. Synthetic imagery has the advantage that the true centers of the spots are known. They can be compared to the results of the fit and lead to a precise quantization of the fit error. The average localization error $\overline{\Delta\mu_{x,y}}$ is the mean error standard deviation and given by Equation (4.50).

$$\overline{\Delta\mu} = \sqrt{\frac{1}{N} \sum_i (\mu_{i,\text{fit}} - \mu_{i,\text{true}})^2} \qquad (4.50)$$

The result of these simulations with synthetic signals, background, and Poisson noise can be seen in Figure 4.14. The average fit error is plotted for the dataflow implementation and the previously used iterative least-square fit. For comparison, the fluoroBancroft algorithm [113] was also included. It provides a noniterative alternative to implementations that are based on the center of mass. For each data point, 2000 spots were simulated.

The fit error is given with respect to the signal-to-noise ratio (SNR). It is defined as in Kubitscheck et al. [122] and given in Equation (4.51). It depends on the background level B, the peak intensity q_{max} of the spot without background, and the Poisson noise level at the peak $\sigma_{q_{max}}$. The localization error is given in units of pixels and units of nanometers for a common pixel area of 102×102 nm.

$$\text{SNR} = \frac{q_{max}}{\sqrt{B + \sigma_{q_{max}}^2}} \qquad \sigma_{q_{max}}^2 = B + q_{max} \qquad (4.51)$$

$$= \frac{q_{max}}{\sqrt{2B + q_{max}}} \qquad (4.52)$$

The charts show that the accuracy of the dataflow implementation is close to the previous least-square implementation. For a low background level with $B = 10$ photons per pixel, the dataflow implementation consistently outperforms the least-square fit for all SNRs by about 5% (Figure 4.14a). The gap narrows for higher SNRs. A background of $B = 10$ can typically be found in images where few parts of the biological sample happen to be distributed outside the focus plane of the microscope.

For a higher background $B = 100$, the least-square fit performs slightly better (Figure 4.14b), especially for a low SNR. The dataflow implementation closes the gap in accuracy for spots with a higher intensity, and both fits asymptotically reach an accuracy of 5 nm or $1/20$ of a pixel.

During development, the effects of zero suppression and spot separation on the least-square fit were also evaluated. These steps improved the quality of the dataflow implementation. For the least-square fit, however, the removal of values from the periphery introduced discontinuities to the signal of the underlying Airy disk and interfered with the convergence of the fit iterations.

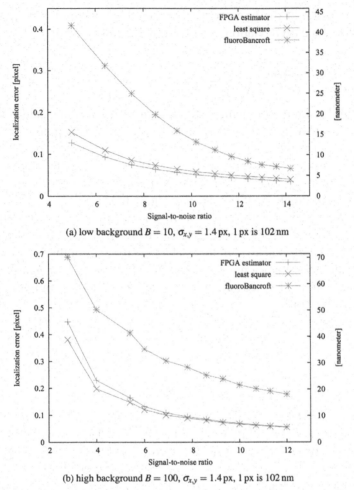

(a) low background $B = 10$, $\sigma_{x,y} = 1.4$ px, 1 px is 102 nm

(b) high background $B = 100$, $\sigma_{x,y} = 1.4$ px, 1 px is 102 nm

Figure 4.14   Monte Carlo simulation of the localization accuracy. The algorithm on hardware (FPGA estimator) is about as accurate as the numerical fits within a 5% margin. For a low background ($B = 10$), it even outperforms the numerical fit. The ROI was chosen 7×7 pixels [92].

The effect of the spot separator can be seen in Figure 4.15. The sample shows tight junctions that form a mesh structure between cells [123]. The recording contained a large number of spots that often touched each other. With the spot separator, some areas of the mesh are more clearly visible, especially at the right border of the image. The need for spot separation was raised by the end users in the research group of Prof. Cremer in Heidelberg

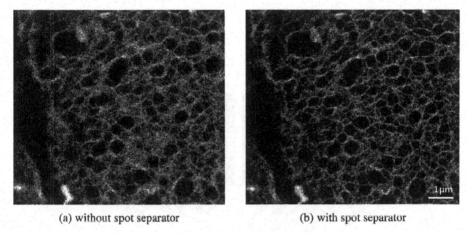

(a) without spot separator   (b) with spot separator

Figure 4.15 Improvement in resolution owed to the signal separator. For image stacks where spots regularly touch each other, the spot separator sharpens the resulting localization image. The mesh structure in the tight junction is less visible without separation.

who observed a clustering of signals along the lines of the mesh that would have been deemed unlikely if only synthetic imagery had been available. We found that the estimated fit error $A\beta_{xy}$ matches the true fit error well if we allow the separator to only remove less than 30% of the total intensity in a ROI and otherwise discard the signal as inseparable.

The localization microscopy challenge at the IEEE International Symposium on Biomedical Imaging 2013 [124] provided an opportunity to compete with other research groups in terms of localization accuracy. The true positions were only known to the organizers of the challenge, who also conducted the analysis. Since the algorithm needed to run on standard hardware, Manfred Kirchgessner created a graphical user interface with the Qt toolkit for the C++ version after he had finished his diploma thesis [48]. The recording was read from the provided TIFF image files, and the program produced a localization image and a list of all locations found.

The challenge contained two data sets with synthetic low-density and high-density imageries. For both sets, our algorithm performed accurate and consistently scored in the upper third of the board. For the high-density image stack "HD3," it achieved the best accuracy among the 15 contestants as shown in Figure 4.16. The tendency seen in the chart indicates that the average localization error increases with the number of points found. In the dataflow implementation, the average accuracy can be traded for a larger number of locations by increasing the threshold for the spot finder.

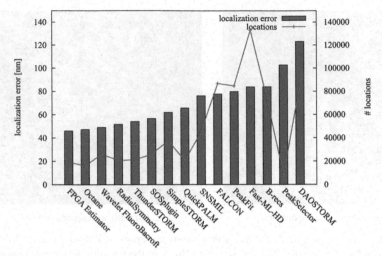

Figure 4.16   Results of the high-density contest at the ISBI localization microscopy challenge [124]. The diagram shows a trend that accuracy degrades with a higher location count. For the sample HD3, our algorithm "FPGA estimator" reached the best accuracy. More locations could have been obtained by increasing the spot threshold, most probably also increasing the localization error.

The accuracy of an algorithm can be also examined with real-world data if the shape of the data is known. The chart in Figure 4.17 displays the profile of a biological structure that forms a line that is thinner than the optical resolution limit. The dataflow implementation shows a profile thinner by 12% than the original iterative least-square algorithm, indicating a smaller deviation from the line. This is largely because the dataflow implementation drops to zero faster in the chart than the least-square fit and therefore produces a higher contrast.

To rule out errors and problems in the algorithm that may not show up during evaluation, but in practice, subsequent software releases for MATLAB on CPUs were created that produce the same results as the dataflow implementation. These releases were then given to other research groups. It was well-received due to the already considerable speedup it brought to these users and provided valuable feedback in return. As a result, the algorithm was not changed, but checks were introduced that warn the user if, for example, the brightness of the light source varied and would have interfered with the exponential smoothing in the part of the algorithm that removes the background. During evaluation, the intensity of the laser was initially assumed to be constant, but real-world usage showed that it cannot be relied on when the user operates a complex microscopy setup.

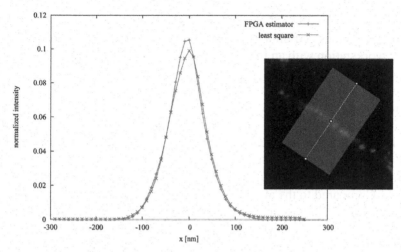

Figure 4.17 Line profile with real-world data. The line is formed by a molecular structure of a cell and is considerably thinner than the diffraction limit of light microscopy. The yellow area indicates the part of the localization image that was analyzed parallel to the yellow line. The results from the Gaussian estimator have a sharper width $\sigma_{\text{estim}} = 41.7$ nm than the least-square method ($\sigma_{\text{lsq}} = 47.5$ nm) [92].

4.1.11.2 Throughput

After the algorithm was redesigned from a least-square fit toward a Gaussian estimator, the original MATLAB program was changed accordingly. In the new implementation, we chose matrix operations over nested loops wherever possible since MATLAB as an otherwise interpreted language is known to take advantage of them [120]. The modification of the software alone in the same environment yielded an acceleration factor of more than 100. On an Intel i5 450 computer with MATLAB 7.10.0, the analysis of a 256×256 pixel image dropped from 7.41 s to 73.9 ms. The input frames could therefore be processed with a throughput of 0.89 Mpx/s. The time used for feature extraction consumed 68% of the computing time and scales with the number of spots found.

The algorithm was also implemented in C++ close to the new MATLAB code. MATLAB can be susceptible to delay code executing in its interpreter. It was found, however, that the C++ version was slower by up to a factor of two and could not compete with the matrix operations in MATLAB that are optimized for CPU vector instructions.

The dataflow implementation was carried out on a MAX2 board from Maxeler Technologies with a Xilinx Virtex-5 LX330T FPGA, where the design achieved a clock frequency of 200 MHz. The board is hosted in a workstation with an Intel Core i5 CPU 750 at 2.67 GHz and 4 GB of RAM. The design was later also synthesized on a MAX3 board for a Xilinx Virtex-6 SX475T inside of a workstation with an Intel Core i7 CPU 870 at 2.93 GHz and 16 GB of RAM.

The runtime of the application can be seen in Figure 4.18 for both systems. Each data point was measured ten times to obtain the mean runtime and its variance. As expected, the runtime rises linearly with the number of input frames. The MAX2 system, however, runs considerably slower with 69.5 MPx/s and has a bigger runtime variance than the MAX3 system. The theoretical maximum throughput is 200 Mps/s, with one pixel consumed per clock cycle. Compared to the already accelerated MATLAB implementation, the MAX2 system achieves an additional acceleration factor of 78, and the MAX3 system speeds up the analysis by 185. In total, an acceleration of 18,500 was achieved when compared to the original MATLAB version.

The first kernel consumes one input pixel per clock cycle, and the feature extraction kernel requires $7 \times 7 = 49$ clock cycles to process the ROI of a spot. For a recording with 2000 frames with 280×320 each and a total of about 300,000 detected spots, the kernel for feature extraction was measured to be occupied for 7.7% of the total time the analysis was running.

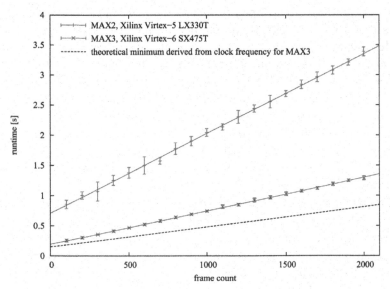

Figure 4.18 The runtime rises linearly with the number of frames with size 288×320 px. The MAX2 board achieved a throughput of only 69.59 ± 0.47 Mpx/s, while the implementation on the MAX3 board runs more streamlined and reaches 167.20 ± 0.59 Mpx/s. The maximum theoretical throughput derived from the clock frequency is 200 Mpx/s.

4.1.11.3 Resource usage

The resource usage for the dataflow implementation on the MAX2 board is shown in Table 4.2. The FPGA is occupied by about 1/3 for FFs and BRAMs, and less than 1/4 for LUTs and DSPs. The pipeline stages for background measurement have the biggest footprint on the BRAM where the background map is stored. Signal separation mainly consists of FFs to implement the comparators and the access pattern from the center toward the border in a ROI. The feature extraction step finally implements the formulas of the Gaussian estimator and consumes the majority of flip-flops for the accumulators and the following floating-point division. All other resources are used for logic to synchronize the kernels and the PCIe interface shared with the host CPU.

On the MAX3 resources, usage was similar (Table 4.3). The main difference was the consumption of 56 DSPs. All other resources were consumed less and stayed within a 10% margin.

After scheduling, the pipeline of the kernel for spot detection consisted of 298 MaxJ nodes and had a latency of 1791 clock cycles. For the kernel for feature extraction, a pipeline with 639 MaxJ nodes and a latency of 280 clock cycles were generated.

Table 4.2 Resource usage on a Xilinx Virtex-5 LX330T FPGA (MAX2) for localization microscopy. The clock frequency is 200 MHz. The resources for synchronization and I/O are included in the total number [92]

Component	LUTs	FFs	BRAMs	DSPs
Background measurement and subtraction	1179	1308	51	0
Spot finding	211	198	0	2
Spot separation	1401	3303	0	0
Feature extraction	23,146	33,817	0	42
Total resource usage	42,044	63,881	108	44
Total resources available	207,360	207,360	324	192
Total resource usage ratio	20%	31%	33%	23%

Table 4.3 Resource usage on a Xilinx Virtex-6 SX475T FPGA (MAX3) for localization microscopy. The clock frequency is 200 MHz. With regard to the Virtex-5 LX330T, 12 additional DSPs were used in total, but fewer LUTs, FFs, and BRAMs

Component	LUTs	FFs	BRAMs	DSPs
Background measurement and subtraction	1301	1343	51	0
Spot finding	414	333	0	2
Spot separation	1257	3369	0	0
Feature extraction	26,969	38,302	0	54
Total resource usage	41,336	62,897	96	56
Total resources available	297,600	595,200	1064	2016
Total resource usage ratio	14%	11%	9%	3%

The source code for the application consists of 1517 lines of MaxJ for the hardware description and 1970 lines of C++ for the host code. For the evaluation of the new algorithm, 3886 lines of MATLAB code were written. The hardware description took about three month to write down after the algorithm was understood and after the new fit method was implemented in MATLAB and C++.

4.1.12 Discussion

The acceleration of the analysis for localization microscopy is a result of a combined algorithm and hardware acceleration approach. Most of the time was spent on rewriting the algorithms while carefully maintaining the same level of accuracy. When the algorithm was brought into a design suitable for pipelining, the following implementation is straightforward with the MaxCompiler libraries.

The results show that the Gaussian estimator with the chosen background removal, zero suppression, and spot separation is as accurate as the compute-intensive least-square fit in MATLAB. This is surprising, given the simplicity of the final algorithm. No iterations are needed and all features of a spot can be calculated within a single scan of the ROI. This made a pipelined implementation feasible that takes one pixel per clock cycle as an input. An implementation of the iterative least-square fit would require the calculation of the fit function (Equation 4.22) and its derivatives with respect to the fit parameters an unknown number of times until a convergence criterion is reached. The calculation of the exponential function would have led to an increased resource usage on the FPGA and a lesser throughput [92].

The change in the algorithm accounted for an acceleration factor of more than 100, and the dataflow implementation in hardware multiplied an additional acceleration factor of 185 on top of it. For the user of the dataflow implementation, this results in an acceleration of 18,500. As we will see for the application for electron tomography, a direct connection of the microscope with the DFE is likely to reach the theoretical maximum of an acceleration factor of 22,500. In particular, with the MAX2 board, a big fraction of the maximum speed is lost due to I/O.

In a compute center, we would not compare the performance of a single CPU core with a DFE, but the performance with regard to rack space. For the CPU, we assume a machine with 24 cores. The MPC-X can be ordered with eight Max3 DFEs, which will also consume just one unit of rack space. With this setup, the dataflow implementation would be about 62 times faster than the improved CPU implementation. At the moment, data sizes do not require a multi-DFE or multi-core implementation, but a larger image stack could be split for parallel processing in the space or time domain to further increase performance.

Compared to the iterative MaLiang method on graphics cards, the speedup is 2.5 for the MAX3 implementation [117]. The difference is rooted in the change of the algorithm, since the FPGA resources were only occupied by less than 1/3. Otherwise, with a similar algorithm, the dataflow implementation should have shown an inferior performance in terms of raw hardware performance. We assume two DFEs per graphics card when comparing performance in the same rack space, leading to an acceleration factor of five.

The throughput of the algorithms allows the processing of raw data in real time for current microscope setups that are usually below 10 Mpx/s, which is more than one order of magnitude slower than the DFE. For most applications of microscopy, the speed of the accelerated MATLAB implementation is already fast enough and a relief for the operators. Since its development, our method has gained traction and is now in use in several biomedical research groups. Along with the ISBI localization microscopy challenge, we can therefore be sure that all major flaws of the algorithms would have been caught and that the usability of the algorithm meets the standards of the research groups involved.

The resource usage indicates that the algorithm would also fit on smaller and cheaper FPGAs such as the Xilinx Artix-7 A105. FPGAs of this size already handle the read-out of some cameras and could be used to perform the analysis for localization microscopy within the housing of the camera in the future.

4.2 3D Electron Tomography

The second of two applications that was accelerated is image reconstruction for computed tomography. A proof of concept was implemented that shows how computed tomography can benefit from a dataflow design on an FPGA and that it performs better than an implementation on a state-of-the-art graphics card. Three-dimensional electron tomography was chosen for the implementation, but the design can also work with other types of radiation.

Computed tomography (CT) aims to reconstruct a density distribution of a sample from a set of projections. It is widely used in medicine and biology to obtain 3D images after the sample was X-rayed from multiple directions and the projections were recorded with a 2D image sensor. Although X-rays are the most common, other types of radiation can be used such as visible light for semitransparent objects or—as for our application—electron rays in an electron microscope.

An overview of the image recording process is shown in Figure 4.19 for electron tomography. Inside of an electron microscope, the volume of the sample is illuminated from one side with electrons (e^-) and a projection is cast on the image detector. After the projection image has been read out, the sample is tilted by a projection angle φ and the next projection is produced, until the sample was illuminated from about 50 directions. For electron tomography, the ray is often not perfectly perpendicular to the tilt axis and will feature a nonzero declination angle α. Depending on the microscope optics, it may also be slightly twisted between sample and electron detector.

The volume is mathematically discretized into cubic voxels for the reconstruction of its density distribution. Due to the mentioned aberrations from a perfectly aligned electron ray, a layer of these voxels will be projected on multiple lines of the detector, even if voxels and detector pixels have the same edge length. Therefore, the problem of reconstructing the density distribution of the sample cannot be divided into a stack of 2D reconstructions from individual lines of detector pixels as it is usually done for X-ray tomography. Reconstruction for electron tomography is harder to parallelize, and its solution already includes a solution for other types of CT.

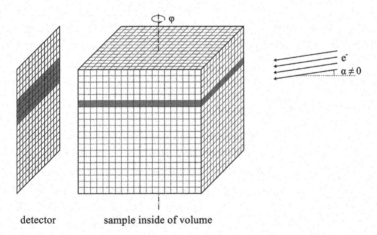

detector   sample inside of volume

Figure 4.19 Overview of an electron tomography recording. The sample inside the discretized volume is projected from multiple angles φ_k to a CCD camera in the electron microscope. A layer of voxels is projected on a band of detector pixels for nonzero declination angles $\alpha < 10°$ or twisted beams [125].

Examples of the capabilities of computer tomography can be seen in Figure 4.20. With electron rays, the resolution of the reconstructed 3D volume approaches 5 nanometers. Electron tomography provides insights into microscopic samples such as biological cells and their organelles and has become an important imaging method for research.

4.2.1 Reconstruction Algorithms

When an electron beam travels through the sample, its intensity is weakened by the density of the material. The decrease is quantified by the detector and is proportional to the integrated density along the path of the beam. If the density of each voxel v_i was known, the intensity decrement p_i on the detector would hold equation 4.53 for each pixel. In the future, we will always assume that p_i is the difference between the unobstructed intensity and the measured intensity for detector pixel i. The sum runs over all voxels, and the weighting factor w_{in} indicates whether and how much voxel n was hit by the ray ending in pixel p_i. For most combinations of rays and voxels, the weighting factor will be zero because the ray missed the voxel.

$$0 = p_i - \sum_{n=1}^{N} w_{in} v_n \qquad (\forall i) \qquad (4.53)$$

The formula is a discretized version of the Radon transformation. It transforms a scalar function defined in a multi-dimensional number space to its projections from every direction and can be exactly inverted. In practice, however, only a

(a) Surprise egg from X-ray images [128]   (b) Eukaryotic cell, electron tomography [129]

Figure 4.20   Examples of 3D reconstruction from 2D projections. Tomography may be known best from computer tomography with X-rays, an example with a surprise egg is shown in (a). Electron tomography uses an electron microscope and has a resolution of about 5 nm. It can be used to reconstruct the architecture of biological cells (b). Here, the actin network of the cell was colored red afterward.

limited number of projections can be obtained from an electron microscope. For these projections, the system of linear equations is underdetermined, because the volume has many more voxels with unknown densities than there are signals available from the detector pixels for projection angles φ. As a consequence, the distribution of densities in the sample cannot be calculated analytically from Equation (4.53) alone.

The algebraic reconstruction technique (ART) starts with an empty volume of voxels and refines the voxel densities iteratively to match the projections. It was invented in 1970 by Gordon et al. [126]. The technique is based on previous works presented in 1937 by Stefan Kaczmarz [127]. The algorithm consists of three steps that are repeated for each voxel in the reconstructed volume, each pixel on the detector, and each projection. It starts with an empty volume.

1. A virtual projection of the volume is calculated for the current pixel. The ray that ends in this pixel is partially absorbed by the voxels of the current iteration. This step is called the (virtual) forward projection.
2. The residue for the pixel is calculated by subtracting the pixel value from the measurement from the virtual one.
3. The voxel densities in the virtual volume along the ray's path are adjusted to match the measured pixel value. The voxel densities are increased by the equally spread residue. This step is called the back projection.

By repeating the process for every pixel i and every measured projection k, the calculated densities in the voxels are expected to approach the real values. The iteration formula is given in Equation (4.54). In the nominator, the difference between measured and calculated forward projection is given for a pixel value pi, and the difference is applied back to the voxels. The relaxation parameter λ is chosen within the range $0 < \lambda \leq 1$. The more accurate the calculation, the closer A can approach 1.

$$v_j^{(k+1,i)} = v_j^{(k,i)} + \lambda w_{ij} \frac{p_i - \sum_{n=1}^{N} w_{in} v_n^{(k,i)}}{\sum_{n=1}^{N} w_{in}^2} \tag{4.54}$$

ART is computationally expensive. Every ray that ends at a detector pixel requires a full iteration. The simultaneous algebraic reconstruction technique (SART) was invented in 1984 by Andersen and Kak [130] to accelerate ART. It computes an entire virtual forward projection for all pixels before it is compared to the measured projection p_{φ_k} at angle φ_k and then projects the difference image back into the volume at once (Equation 4.55).

$$v_j^{(k+1)} = v_j^{(k)} + \lambda \frac{\sum_{p_i \in P_{\varphi_k}} \left(\frac{p_i - \sum_{n=1}^{N} w_{in} v_n^{(k)}}{\sum_{n=1}^{N} w_{in}} \right) w_{ij}}{\sum_{p_i \in P_{\varphi_k}} w_{ij}} \tag{4.55}$$

As before, the difference between virtual forward projection and measured projection is calculated in the inner fraction, while the outer parts describe the back projection for all pixels for tilt angle φ. Note that the factor w_{ij} is zero for most combinations of ray i and voxel j, since most voxels do not shadow the given pixel. For an implementation on a computer, both forward and back projections are calculated by following a given ray through the volume and by integrating the voxel densities along its path, skipping the calculation of most zero w_{ij}. Alternatively, the volume can be read linearly voxel by voxel and the densities are accumulated for the shadowed pixels during projection.

The rays for SART do not have to be parallel; they all may also stem from the same point or can have a different configuration. For electron tomography, we will assume parallel beams.

4.2.2 State of the Art

The first implementations of computer tomography were two-dimensional reconstructions and consisted of back projections only. A sample was imaged with X-rays on a one-dimensional sensor for multiple angles. The sample was then reconstructed by accumulating the back projections on top of each other. Since the virtual forward projection is omitted, all back projections are applied with the same magnitude. For a point-like density distribution in the sample, the reconstruction shows a spot that is blurred with a starred shape (Figure 4.21). The star has two spikes for each direction of projection.

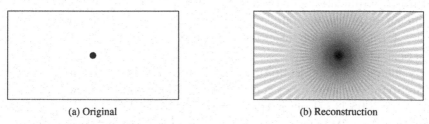

(a) Original   (b) Reconstruction

Figure 4.21 Reconstruction with back projection only. A point-like density distribution was reconstructed by projecting back its (few) measured projections. The circular object in the middle is imaged with a star-shaped artifact and one line for each projection in the corresponding direction.

The technique of *filtered back projection* was invented to mitigate the blurring effect caused by the summed back projections. A high-pass filter is applied on the reconstructed image and sharpens the reconstruction. The projection slice theorem [131] allows the application of the filter on the one-dimensional recordings for the same effect and greatly reduces the computational power needed. The filtered back projection works especially well if many projections can be recorded. This is usually the case for X-ray tomography, but not for electron tomography. Filtered back projection can be easily computed in parallel with one thread for each projection. For 2D images with 1000×1000 pixels and more than 1000 projections, the 2D image can be reconstructed in the order of a second on state-of-the-art devices [132].

For tomography with a lesser number of measured projections available, ART and SART proved to reconstruct with a higher accuracy. The total number of projections in electron tomography must be balanced between reconstruction quality that benefits from an increase in projections and radiation damage of the sample that must be kept low. The need for processing power is further tightened by improvements of the resolution of the CCD sensors in the microscope. The SPIDER and WEB software packages have been developed in 1996 to incorporate multi-core processing into reconstruction for transmission electron tomography [133]. Later in 2004, the Xmipp software package further accelerated reconstruction by cluster computing for Unix-like systems with full data format compatibility [134].

Further acceleration was achieved with graphics cards after they started to support floating-point arithmetic natively. Daniel C. Díez et al. showed in 2006 that SART and related techniques can be sped up by a factor of sixty to eighty with commercially available graphics cards [135] compared to the performance of a single CPU. A machine with a single graphics card could therefore substitute an entire medium-range computing cluster. With further optimizations of the reconstruction algorithm, an extra factor of about an order of magnitude could be achieved in 2010 on similar hardware for transmission electron tomography with parallel beams by Wei Xu et al. [136].

Only filtered back projection has been implemented on reconfigurable hardware so far. Miriam Leeser et al. ported the algorithm for parallel beams of X-rays in 2005 [137] on a Xilinx Virtex 2000E FPGA and achieved a speedup of 100 for 2D reconstruction. SART encompasses the calculations of forward and backward projections and was not implemented on reconfigurable hardware yet.

4.2.3 Analysis of the Algorithm

In order to implement the algorithm as a system of pipelines, we must find a dataflow description that fits the constraints of the DFE and its periphery. We started with an implementation on the graphics card given by the research group of Prof. Frangakis, an enhancement of the algorithm presented by Daniel C. Díez et al. [135] that was optimized for the architecture of modern graphics cards. It calculates forward and back projections by following each ray through the volume. The density of the volume is sampled along the ray in intervals of one third of the voxel edge length for integration. The volume is kept in the memory of the graphics card and therefore was restricted to about 2 GB at the time of development. For the forward projections, the voxel densities are integrated along the path of the ray. After comparison with the measured projections, the residue is projected back by spreading it equally on each voxel along the path. On the DFE, the voxel densities cannot be kept in the few megabytes available with BRAM storage, and therefore, the volume has to be stored in the onboard DRAM.

The location of the volume storage in the DRAM leads us immediately to consider the access speed of the voxel densities. The transfer frequency of DRAM is limited to less than 143 MHz for random access and can only be read faster in a sequential order (Section 3.5). If we aimed for a hardware pipeline that processes one voxel per clock cycle, the pipeline would stall below 143 MHz, which is already at the low end of the capabilities of a modern FPGA. In case, the number of FPGA resources allows for the design to be multi-piped, the limitation of random DRAM access becomes a bottleneck that prevents further acceleration.

Figure 4.22 sketches in which order the voxels have to be accessed in the graphics card implementation. Following a ray through the volume, the arising access pattern is pseudo-random instead of linear. The direction of the rays changes with each change of recorded projection and can therefore not be globally re-arranged in the direction of the rays. Re-arrangement between projections, however, would introduce pseudo-random access again.

4.2.3.1 Modifications

The solution to the access limitation is to rewrite the algorithm to process the volume voxel by voxel instead of ray after ray. For each voxel, the rays that are likely to hit it have to be found and tested for intersection. The virtual projection can be built by accumulating the pixel values for each ray separately, until the entire volume was read. In summary, the random

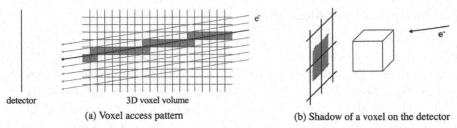

(a) Voxel access pattern   (b) Shadow of a voxel on the detector

Figure 4.22   Pseudo-random access pattern of voxels along a ray of electrons (e^-). The graphics card version of the algorithms follows ray after ray and samples the voxel density three times per edge length (red markers). This creates a pseudo-random access pattern in DRAM (blue). For maximum throughput, DRAM should be read sequentially voxel by voxel in bursts of data instead. A voxel at position $\vec{v}$ casts a shadow on up to four detector pixels with positions $\vec{p}_i$.

access to the volume is traded for random access in the storage for the virtual projection. Because the projection is two-dimensional, its size only grows with the power of two of the volume edge length and fits into the BRAM storage, while the volume grows with the power of three and must be kept in DRAM. For the projection in BRAM, random access is equally fast as sequential access.

A given voxel can shadow up to four detector pixels for the common case that pixels and voxels share the same edge length (Figure 4.22b). The corresponding four rays are found by projecting the center of the voxel onto the detector and selecting the four closest pixels. The centers of the pixels then give the endpoints of rays to be tested for intersection. If ray and voxel intersect, the intersection length will be given by the distance between the point where the ray enters the voxel and the point where it leaves the voxel again. A voxel may only be hit by up to four rays, but a ray hits every voxel on its path and is cumulatively weakened by every voxel with a nonzero density. The length of the ray–voxel intersections has to be computed for both forward and back projections and must be repeated for every voxel for every projection.

The four detector pixels shadowed by a voxel are found mathematically by projecting the center of the voxel onto the detector. The direction of the electron rays can be identified from the experiment by knowing the the position of the current voxel and the orientation of the detector. Using an affine projection, the voxel can be transformed into the coordinate system of the detector and projected onto it. An affine transformation contains a linear matrix operation such as a rotation, and a translation by a fixed amount. The formula of the projection is shown in Equation (4.56) with v as the position of the voxel, M as a 2×4 projection matrix, and $\vec{p}$ as the position on the detector. It requires six multiplications and four additions in total. Here, the first three rows of M describe the linear transformation of v and the last row contains the translation vector.

$$\begin{pmatrix} p_x \\ p_y \end{pmatrix} = \begin{pmatrix} M_{11} & M_{12} & M_{13} & M_{14} \\ M_{21} & M_{22} & M_{23} & M_{24} \end{pmatrix} \begin{pmatrix} v_x \\ v_y \\ v_z \\ 1 \end{pmatrix} \tag{4.56}$$

By using units of the detector pixels and rounding the x- and y-coordinate of $\vec{p}$ up ($\lceil \cdot \rceil$) and down ($\lfloor \cdot \rfloor$) to the nearest integer, we obtain the coordinates of the four closest detector pixels $\vec{p}_1, \vec{p}_2, \vec{p}_3$ and $\vec{p}_4$. The coordinates of these pixels are then equal to the endpoints of the four rays that need to be tested for intersection with the current voxel (Equation 4.57). Please observe that the components of the vectors differ only by one unit in a direction, for example, $\vec{p}_2 = \vec{p}_1 + \vec{e}_y$.

$$\vec{p}_1 = \begin{pmatrix} \lfloor p_x \rfloor \\ \lfloor p_y \rfloor \end{pmatrix} \quad \vec{p}_2 = \begin{pmatrix} \lfloor p_x \rfloor \\ \lceil p_y \rceil \end{pmatrix} \quad \vec{p}_3 = \begin{pmatrix} \lceil p_x \rceil \\ \lfloor p_y \rfloor \end{pmatrix} \quad \vec{p}_4 = \begin{pmatrix} \lceil p_x \rceil \\ \lceil p_y \rceil \end{pmatrix} \tag{4.57}$$

After the endpoints of the four rays have been determined, they need to be transformed into the coordinate system aligned with the volume. This requires another affine transformation with a 3×3 matrix on each endpoint before the rays can be tested for intersection (Equation 4.58).

$$\vec{p}_i' = \begin{pmatrix} p_x' \\ p_y' \\ p_z' \end{pmatrix}_i = \begin{pmatrix} N_{11} & N_{12} & N_{13} \\ N_{21} & N_{22} & N_{23} \\ N_{31} & N_{32} & N_{33} \end{pmatrix} \begin{pmatrix} p_x \\ p_y \\ 1 \end{pmatrix}_i \qquad i \in \{1,2,3,4\} \tag{4.58}$$

One of these transformations requires six additions and six multiplications. For all four pixel positions $\vec{p}_i'$, however, the cost does not rise by a factor of four. The x- and y-coordinates of $\vec{p}_i'$ have been rounded to integer values in units of pixels before and differ only by one. After the transformation for position $\vec{p}_1'$ with both components rounded down has been computed, the other vectors are obtained more efficiently by adding either the matrix columns (N_{11}, N_{21}, N_{31}), (N_{12}, N_{22}, N_{32}), or both on top of $\vec{p}_0'$ (Equation 4.59). Hence, no further multiplications are introduced and a total of 15 additions are needed.

$$\vec{p}_2' = \vec{p}_1' + \begin{pmatrix} N_{12} \\ N_{22} \\ N_{32} \end{pmatrix} \qquad \vec{p}_3' = \vec{p}_1' + \begin{pmatrix} N_{11} \\ N_{21} \\ N_{31} \end{pmatrix}$$

$$\vec{p}_4' = \vec{p}_1' + \begin{pmatrix} N_{12} \\ N_{22} \\ N_{32} \end{pmatrix} + \begin{pmatrix} N_{11} \\ N_{21} \\ N_{31} \end{pmatrix} \tag{4.59}$$

The matrices M and N depend on the geometry of the microscope and can be computed on the host computer. They only change when a new iteration starts after a projection has been completed.

The graphics card implementation samples the volume along the ray in intervals of one third of the voxel edge length. For the dataflow implementation, which operates voxel by voxel, we cannot use this method. The voxel is already known and the rays have to be found instead. To obtain the intersection lengths of the four rays and the voxel, we use four instances of *Smit's algorithm* for ray–box intersection [138, 139]. Here, the box is the voxel in the coordinate system aligned with its edges. The algorithm was adapted from the original description to handle rays that are not defined by the origin, as it is normally the case, but by their endpoint. Its description is given in Equations (4.60) to (4.66) and illustrated in Figure 4.23.

$$\vec{t}_{\text{entry}} = (\vec{v}_{\text{min}} - \vec{p}')/\vec{r} \tag{4.60}$$

$$\vec{t}_{\text{exit}} = (\vec{v}_{\text{max}} - \vec{p})/\vec{r} \tag{4.61}$$

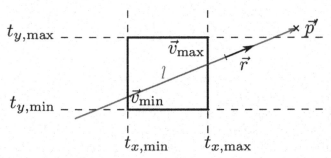

Figure 4.23 Ray–voxel intersection. The densities of the voxels are integrated along the path of the ray. To calculate the intersection length l (red), all coordinates are first transformed into the coordinate system aligned with the voxel (black box). Afterward, the intersections t of the ray with the layers of the voxel are determined as multiples of $\vec{r}$, sorted, and subtracted (Equations 4.60 to 4.66). The drawing is reduced to two dimensions for clarity.

$$\vec{t}_{\min} = \vec{\min}(\vec{t}_{\text{entry}}, \vec{t}_{\text{exit}}) \tag{4.62}$$

$$\vec{t}_{\max} = \vec{\max}(\vec{t}_{\text{entry}}, \vec{t}_{\text{exit}}) \tag{4.63}$$

$$t_{\text{in}} = \max(t_{\min,x}, \max(t_{\min,y}, t_{\min,z})) \tag{4.64}$$

$$t_{\text{out}} = \min(t_{\max,x}, \min(t_{\max,y}, t_{\max,z})) \tag{4.65}$$

$$l = \max(t_{\text{out}} - t_{\text{in}}, 0) \tag{4.66}$$

The position and dimensions of the voxel are given by its opposing corners $\vec{v}$ min and $\vec{v}$ max, where $\vec{v}$ min points to the corner with the numerically smallest vector components. The ray is given by the endpoint $\vec{p}'$ on the detector and its normalized direction $\vec{r}$. The quotient $1/\vec{r}$ and the three-dimensional minimum function $\vec{\min}()$ and maximum function $\vec{\max}()$ are calculated component by component.

The ray–voxel intersection length l is computed by first calculating the intersection of the ray with the six layers that border the voxel. The positions of the respective six intersection points are stored in $\vec{t}_{\min}$ and $\vec{t}_{\max}$ as multiples of the direction r, starting from the endpoint $\vec{p}'$. $t_{\min}$ and $t_{\max}$ also contain the information where the ray hits the voxel, but must be sorted first in Equations (4.62) and (4.63) to not confuse entry and exit. The position of the entry of the ray into the voxel is then given by $p' + t_{\text{in}}\ \vec{r}$, and similar for the exit with t_{out}. Since $\vec{r}$ is normalized to a length of one, the intersection length is the difference of t_{out}. minus t_{in}. If the ray misses the voxel, the difference will be negative. In this case, it is set to zero in Equation (4.66).

Note that every variable is assigned only once. This makes the computation immediately transferable to a synchronous pipeline in dataflow computing.

4.2.3.2 Dataflow

We can immediately infer a circular dataflow from the iterative nature of the algorithm. The currently reconstructed volume is projected onto a virtual projector, the measured projection is compared with the virtual projection, and the difference is projected back into the reconstructed volume. Then, the next iteration continues. The high-level dataflow can be seen in Figure 4.24. It also gives a first hint for the implementation.

At runtime, the design starts with an empty volume in the DRAM. It is read linearly voxel by voxel and fed into logical block A, where the four rays that may hit the current voxel are determined. A hit consists of the intersection of ray and voxel, and its length corresponds with the weights w_{ij} in Equation (4.55). In block B, these hits are multiplied with the voxel density and accumulated for each pixel they shadow on the virtual detector. Here, the

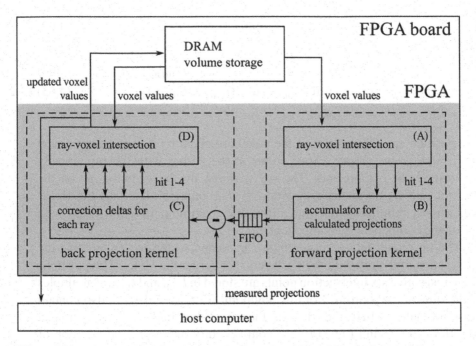

Figure 4.24   High-level dataflow between hardware components. The iterations of the reconstructed volume are kept in DRAM. Its voxels are piped through the DFE, where the ray–voxel intersections (A) are computed to accumulate the forward projection (B). After the forward projection has been obtained, it is compared with the measured projection and the difference (C) is projected back into the volume (D) [125].

virtual projection is gradually generated line by line. When a detector line has been fully computed, it is compared with the measured projection from the host computer. The residues are stored in block C for lookup and are required for the back projection. The last element of the circular dataflow is again a test for the ray–voxel intersection in block D. This time, however, the residues are read from block C and applied onto the reconstructed voxel density from the beginning of the iteration. An iteration for one projection is done when all voxels have been processed and the next projection can be started. In the last iteration, the voxel values are sent to the host computer and constitute the final reconstruction. Blocks A and B form the logic for the forward projection, while blocks C and D apply the back projection onto the volume.

The dataflow allows for all stages to operate in parallel most of the time. The only delay occurs if block B has not yet accumulated the shadows of all voxels that belong to the first line. In this case, block C must wait until the line is completely computed and can be compared with the measured projection. The higher the declination angle, the more layers of voxels will contribute to a line on the detector and the longer the back projection in blocks C and D will have to wait (see also Figure 4.19). A small declination angle close to zero is therefore desired for maximum performance. We will see later in the implementation section how to decouple forward and back projection to account for their nontrivial synchronization.

All operations in the dataflow operate on the stream of voxel values, each forming one data item to process in the pipeline. For a 3D volume with edge length L, the number of operations on voxels is of order $\mathcal{O}(L^3)$, while the number of operations on pixels in a projection is only of order $\mathcal{O}(L^2)$. Hence, with L up to 1000, all operations that run on voxels are the most time-intensive and should be ported to the DFE, while operations on the values of the detector pixels may stay on the CPU.

4.2.3.3 Dimensioning of the hardware

The knowledge about the dataflow already gives us a description close to the actual implementation of the hardware. The last parts missing from the high-level design are the type of parallelism to use and the data types for the multiple variables.

Parallelism: The dataflow graph shown in Figure 4.24 suggests that we can process all data with pipeline parallelism, continuously streaming the reconstructed voxel densities from and to the volume storage in the DRAM. For this, the individual algorithmic blocks must fit into the same streaming model of one data item per clock cycle. We will start with a single pipeline that processes one voxel per clock cycle (MISD) and later extend the design with multi-pipeline parallelism (MIMD).

For ray–voxel intersection, we have already seen that the algorithm adapted from computer graphics is of linear nature without conditionals or loops. It can therefore be expected to be implemented with a pipeline that accepts one voxel and ray per clock cycle and emits the length of the ray–voxel intersection. The matrix multiplications needed before are also small enough to be implemented into a pipeline directly without the need for a loop over the vector indexes. These matrix multiplications encompass the projection of the voxel onto the detector that are needed to find the four rays to possibly hit the voxel, and the transformation from the coordinate system of the detector to the coordinate system of the voxel.

The virtual projection is accumulated in BRAM. Here, the voxel densities have to be integrated along the path of the rays through the volume. After the ray–voxel intersection was computed, we need to add the intersection lengths multiplied by the voxel density on top of the previous value of the affected virtual detector pixels to build up the forward projection. The pixel values are kept at a given storage location in dual-port BRAM, where they can be read and written at the same clock cycle. However, accumulating values in BRAM is complicated by the fact that BRAM introduces a latency between reading a value and writing it back. To increase a pixel value, it has to be read first, then incremented and finally written back. For pipelined access, this introduces a read-after-write conflict that leads to errors when the same storage location is incremented continuously for several clock cycles. Fortunately, the access pattern on the virtual detector in the DRAM allows us to avoid this problem with few extra hardware resources: The voxels are linearly read from the volume storage in BRAM, and after the affine transformations have been applied, the access pattern to the pixels in the BRAM is still largely monotonic. The details can be found in the implementation Section 4.2.4.

After the virtual projection was completely accumulated and subtracted from the measured projection, the residue storage is written only once per projection. After this, access remains read only until the next projection, avoiding any read-after-write conflicts and preserving data consistency. The residues are read during back projection when the ray–voxel intersections are computed again and the residues are applied to the volume.

The voxels in the volume are read twice per iteration from the DRAM, first during forward projection and a second time when the density values are updated during back projection. Both times the same ray–voxel intersection lengths are computed. Instead of computing the intersection twice, it would be computationally less intense to store them during forward projection and merely retrieve them during back projection. However, due to the time offset between forward and back projections, the amount of storage needed for the intersection length would require us to store them in onboard DRAM. Read and write access to this intersection buffer would then more than double the total memory access load. A bandwidth of 2.4 GB/s is already required with about 4 bytes for the voxel density, two reads and one write access per clock cycle for forward and back projections, and a target clock frequency above 150 MHz. The remaining DRAM bandwidth was instead kept free for multi-piping the design. The cost for this decision is an increase in resource usage for flip-flops, lookup tables, and DSPs.

Numerics: The dimensions of the number types in the entire design are largely given by the accuracy of the detector in the electron microscope and the size of the volume. The CCD sensor produces a 16-bit integer value for each pixel. The accuracy of a calculated voxel density is then determined by examining the back projection where the densities are reconstructed. Here, the error of ± 1 for the residue is spread equally to all voxels on the path of each ray. The length of a ray is at maximum $\sqrt{2} \times 1024$ for a volume with $2^{10} = 1024$ voxels in each direction and a diagonal ray through the volume assuming a low declination angle. Hence, the accuracy of a voxel density is at most about 2^{-13} with a 2-bit safety margin.

The positions of the (virtual) detector pixels and the reconstructed voxels can be addressed with integer indexes. For an edge length of 1024, a 10-bit integer is sufficient. For the affine projections, the determination of the accuracy is more difficult: The ray–box intersection is sensible to small errors, as the intersection of a ray that barely touches a voxel must be precisely calculated to not produce false negatives or positives. In particular, rays where one component of the vector direction vanishes and the direction becomes parallel to one side of the voxel volume depend on a precise calculation of their endpoint, as a small change in its position makes the difference between a hit with a maximum intersection length equal or greater than one voxel edge length, and a miss with zero interaction [139]. In this case, the intersection length changes from a continuous function of the endpoint position to a discrete one.

To find the right accuracy for fixed-point number encoding, a simulation was carried out that compared the accuracy of the ray–box intersection length using a fixed-point number encoding with the intersection length obtained with a floating-point number encoding. The ray endpoint was randomly chosen with a distance of up to 2048 voxel, the voxel position was also randomly set, and the ray direction was set with a random distribution such that it would hit the voxel with a chance of about 50%. The simulation avoided the case where the ray direction would contain a component with an absolute value of less than 0.0005. In the unlikely event that a vector component of the direction happens to be smaller than that, it will be set to 0.0005. The magnitude of the errors introduced hereby will be given in the results section.

The numeric encoding chosen for the remaining parts of the design is given in Table 4.4. Most encodings are fixed-point encodings with a total number of 32 bits. When used for inputs or outputs, the host code will convert them between floating point for the CPU and packed 32-bit integers for the PCIe bus. The only notable exception from this rule is the core algorithm for ray–voxel

Table 4.4   Numerical encoding used for the algorithms for electron tomography on hardware. All internal operations can be executed with signed fixed-point encodings, leading to a smaller resource footprint than floating point

Use Case	Number Encoding
Voxel density	Fixed-point encoding with 13 integer, 19 fractional bits
Voxel density read-out	Floating-point encoding with single precision
Transformation matrix	Fixed-point encoding with 12 integer, 20 fractional bits
Ray–box intersection	Fixed-point encoding with 20 integer, 30 fractional bits
Intersection length	Fixed-point encoding with 12 integer, 20 fractional bits
Physical detector pixel	Unsigned 32-bit integer
Virtual detector pixel, residue	Signed fixed-point encoding with 16 integer, 12 fractional bits

intersection, which is computed with a 50-bit fixed-point encoding with 30 fractional bits. The results of the analysis that lead to this encoding are shown in Section 4.2.5. Furthermore, the density of the volume is read out during the final iteration encoded as floating-point numbers with single precision.

The use of fixed-point numbers limits the resource usage. The problem of reconstruction features a limited range for the numbers involved, mainly because all operations are related to the limited range of the CCD sensor. The only exception is the ray–voxel intersection that requires the inverse of ray direction to be computed, with the potential of producing arbitrarily large numbers. Since the inverse is precomputed on the CPU for each iteration, we can calculate the inverse components of the ray direction on the CPU with double precision and convert them to fixed-point numbers afterward, limiting the maximum error to a small and well-known magnitude that is smaller than the precision of the microscope.

4.2.4 Implementation

MaxCompiler 2011.3 was used for the description of the dataflow graph and the generation of the intermediate VHDL representation. The bitfile containing the configuration for the FPGA was then synthesized with the Xilinx Integrated Software Environment (ISE) 13.3 for a Xilinx Virtex-6 SX475T FPGA. On the FPGA board, 24 GB of RAM are available, equally distributed on six SO-DIMMs.

The discussion of the dataflow in the previous subsection already gave us a high-level description of the hardware design. The functionality is divided into two domains: the first one for forward projection and the second one for back projection. For the implementation, these domains must be mapped to one or more pipelines in hardware. Backward projection can only start after the first lines of the virtual projection have been completed. Due to a nonzero declination angle of up to 10°, the calculation of the first line of the virtual detection may require the traversal of multiple layers in the volume. Hence, the statically scheduled pipeline of the design has to be split between forward and back projections and must be decoupled with an (elastic) FIFO in between. As a result, the design is made of two kernels in the MaxCompiler source code that are dynamically scheduled by the manager. This design decision is already reflected in Figure 4.24.

4.2.4.1 Scheduling

The scheduling between kernels, DRAM, and the host CPU is done automatically by the manager and the Linux kernel module of MaxCompiler. The kernels run when all enabled kernel inputs are nonempty and the outputs do not stall. Within the kernel pipelines, however, the orchestration of the dataflow is done manually. Contrary to the application for localization microscopy, we cannot resort to the principle of one data item for every clock cycle. The BRAM implementation of the projection accumulator in the forward projection kernel cannot transfer data to the back projection kernel at the same time. A parallel read and write access is needed to increase a value in the virtual detector, and both ports of the BRAM are therefore already occupied. The forward projection must be stopped to transfer the values (read access) and to zero the storage for future use (write access) to transfer the pixels of the virtual detector, in the virtual detector, and both ports of the BRAM are therefore already occupied. The forward projection must be stopped to transfer the values (read access) and to zero the storage for future use (write access) to transfer the pixels of the virtual detector.

The virtual projection has far less pixels than the volume has voxels. For a volume with 1024^3 voxels and a detector with 1024^2 pixels, the forward projection has to be halted for transfer for less than 1‰. The specific implementation of the BRAM accumulator allows to transfer multiple pixel values per clock cycle and reduces the downtime of the forward projection even further.

The position of the window of virtual pixels and residues that have to be kept in BRAM slides monotonically with every new layer of voxels. When a layer of voxels has been processed, the shadow of the next layer advances downward by at most one line of pixels on the detector. It is therefore sufficient to dedicate transfer time for just one detector line after a voxel layer was completed.

The control logic was implemented as a finite-state machine (FSM). An FSM deviates from the dataflow model, but simplifies the formulation of the logic. It is integrated in the statically scheduled pipeline of the forward projection kernel and transitions its state every clock cycle when the kernel is running. The state transition graph is shown in Figure 4.25. It consists of one state where the forward projection takes places, and two states for the transfer of the previously computed virtual detector pixels.

The volume is projected voxel by voxel onto the virtual detector during the initial state "project volume." The state is kept until an entire layer of the volume was traversed, and the index of the voxel is increased to track the position of the voxels. After one layer was projected, the FSM transitions into the state "line transfer." Here, the projection is paused to allow the transmission of a detector line to the back projection kernel. The transmission will only take place if the BRAM accumulator indicates that the calculation of the line was completed. Afterward, the FSM will return to the projection state.

At the end of a projection, when all voxels of the volume have been traversed, the remaining lines of pixels in the virtual detector have to be transferred to the back projection kernel. This is done in the state "final transfer." After the last line has been transferred, the state returns to "project volume" for the next iteration unless the projection was the last one. In the

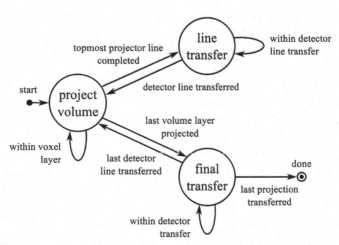

Figure 4.25 Forward projection FSM. The first state projects a layer of the volume onto the virtual detector ("project volume"). When done, the virtual projection is halted and one line of the virtual detector is transferred to the kernel for back projection ("line transfer"). When the entire volume was projected, the remaining lines of the detector are transferred ("final transfer") and the next projection is started.

latter case, the FSM switches into the "done" state, where it remains until the DFE is reset.

The transition graph of the FSM for the back projection kernel is shown in Figure 4.26. Before back projection can start, all residues needed for the first layer of voxels must be available. Depending on the declination angle of the electron microscope, the shadow of the first voxel layer may fill the entire BRAM. During the first "initial projection," the forward projection of the first layer is received and the residues are calculated accordingly by subtracting the real from the virtual projection. The amount of data needed is calculated by the host CPU and called the "initial detector window." As soon as the data are transferred, back projection can start for the first layer of voxels in the state "project volume." When the back projection was applied for the layer and, if needed, the FSM halts back projection and performs the transfer for the next line of residues in the state "line transfer." After the last layer has been projected, the state "final transfer" takes care of consuming any detector pixels from the forward projection kernel that have not been transferred yet between kernels and could otherwise stall the forward projection kernel. When done, the next back projection is ready to start and the FSM continues with the initial transfer for it. Finally, after the last iteration was projected back, the FSM stops and stays in the "done" state until the DFE is reset.

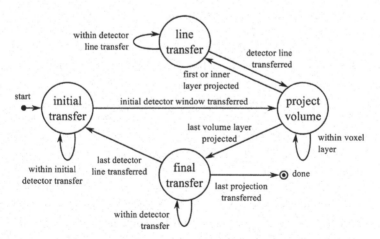

Figure 4.26 Back projection FSM. The first state blocks until the residues needed for the first voxel layer have been received ("initial transfer"). When done, the residues are projected back into the layers of the volume ("project volume"). After a layer was projected back, projection is halted and the next line of residues is received ("line transfer"). At the end of a projection, the possibly remaining residues are consumed by the kernel ("final transfer").

The FSMs for the forward and back projection kernels control all other elements of the hardware. They keep track of the current voxel to be projected forward or back, as well as the pixel transfer between kernels. They also control the sliding windows where the forward projection is accumulated within, and the residues are obtained from for back projection. Each FSM needs to match the behavior of the other one in order to prevent the connecting FIFO from underflows or overflows whenever possible.

When a projection was finished, the FSMs also enable the inputs of both kernels that receive the properties of the next projection for one clock cycle. These are the matrices for the affine transformations and the inverse direction of the electron rays. The data are precomputed on the host CPU and sent to both kernels via PCIe.

4.2.4.2 External DRAM

The hardware design accesses the volume in the DRAM storage through three channels. The first kernel reads the voxel densities linearly to calculate the forward projection. The second kernel updates the voxel densities during back projection and needs to first read each voxel and then write the new value. The CPU does not have read or write access. Instead, the forward projection kernel treats all (uninitialized) densities as zero during the first iteration, and the back projection kernel sends the resulting voxels to the host CPU during the last iteration through an extra output using the PCIe bus.

All accesses to the DRAM are linearly and allow for burst access. The three address generators that produce the access pattern for the DRAM controllers can be instantiated in the Java class that describes the manager. Here, the DRAM and the kernels are instantiated and connected with each other and the PCIe bus. Thanks to the DRAM burst mode, the lowest available DRAM frequency of 303 MHz could be chosen to ease the requirements for timing closure during hardware synthesis.

The linear access pattern compromises a range in the address space and the number of blocks to be accessed. These values are set during runtime by the host code on the CPU before the DFE is started. The corresponding C function is provided as a library call by the Maxeler RT library. After that, the address generators run on its own on the DFE. Their speed is controlled by the throughput of the connected kernels.

The size of a read or write burst is 384 bytes because the hardware combines the bursts of six DRAM SO-DIMMs for maximum performance. The width of the inputs and outputs of the kernels that are connected to the DRAM must be an integer divisor of the burst length to save hardware resources. The storage format was chosen to be a fixed-point encoding with 32 bits, which came closest above the theoretical precision.

4.2.4.3 Ray–Box intersection

The implementation of the algorithm for ray–box intersection is very close to the description given in Equations (4.60) to (4.66), which is already in a format suitable for dataflow computing. The respective accuracies for the fixed-point numbers are listed in Table 4.4 in the previous subsection.

In very rare cases, a voxel may be wrongfully missed by all four rays due to rounding errors. This can only happen if the rays are parallel to one of the voxel's sides and the voxel happens to be exactly in between two pairs of rays. The error introduced for the calculation of the forward projection can be neglected, since a ray traverses through many voxels. For the back projection, this miss can introduce high frequency noise in the volume. The partial pipeline for ray–box intersection therefore checks for a vanishing intersection length for all four rays. In this case, a rounding error caused all four rays to miss the voxel, and the individual intersection lengths are all set to 1/4.

During development, it became apparent that this part of both kernel pipelines is responsible most of the time when the design did not reach timing closure in hardware synthesis. In this part of the pipeline, the encodings for fixed-point numbers contain the largest number of bits and make the multipliers expensive in terms of hardware resources and timing constraints. Overall and as shown later in the results chapter, most DSPs in the design are required for this part of the algorithm. The MaxJ description was carefully tuned with extra pipeline registers to retime the connecting lines between the DSPs. To reach the highest clock frequency for both kernels, every pipeline stage had to be extended with one extra register. The pipeline stage before the multiplication with the inverse ray direction in Equations (4.60) and (4.61) had to be retimed with even two extra layers of pipeline registers.

4.2.4.4 Projection accumulator

After calculating the intersection with four rays for each voxel, the forward projection kernel multiplies them with the voxel density and accumulates the resulting shadow where it is casted on the virtual detector. After a line has received all shadows, its values are transferred to the back projection kernel, where the line of pixels is compared to the real projection obtained from the microscope. The traversal of the volume is chosen such that first the voxels within a layer are visited before the next layer is accessed (Figure 4.19).

Hence, a layer of voxels shadows multiple lines of the detector, but not the entire detector, allowing us to keep only the affected detector lines in BRAM on the FPGA.

For a volume with 1024^3 voxels, the shadow of a layer on the detector fills a band with a width of up to $1024\sqrt{2}\sin(\alpha)$ pixels. For a maximum beam declination angle $\alpha \leq 10°$, this means that we have to keep at least 252 detector lines or about 1 MB in the BRAM. Storing the entire virtual detector would simplify the hardware, but is not an option: The storage needed would amount to 3.6 MB, which is more than half the entire BRAM available on a Xilinx Virtex-6 SX475T FPGA [140], and the same amount would be needed again to store the residues in the back projection kernel. Therefore, only the currently shadowed lines are buffered.

The design of the projection accumulator consists of three layers of hardware. The topmost layer manages the sliding band of shadowed pixels when the volume is traversed voxel by voxel. It also contains the logic to transfer every completely calculated line of the detector to the back projection kernel. The middle layer reads and writes to this storage and adds the received voxel shadows on top of the previously stored pixel value. Finally, the bottommost layer provides the storage needed to accumulate the shadows of the volume.

The topmost layer receives the individual shadows from the four rays that can hit the current voxel. The pixels where these rays end are neighbors and form a 2×2 square on the virtual detector. The memory was therefore split into four parts, one part for each combination of odd and even index for detector column and line (Figure 4.22b). Each part then updates one pixel per clock cycle and acts as its own accumulator. The storage location is determined by using the line number modulo the number of buffer lines. This yields a ring buffer that just contains the detector lines that are shadowed at the moment by the current volume layer, plus some extra lines for the transfer. After a layer of the volume was traversed, it is checked whether a pixel line was left unmodified and is ready to leave the ring buffer during the following "line transfer" state of the forward projection FSM. The line is then zeroed and can be used again for a different line index.

The middle layer contains the accumulation logic and is instantiated four times per voxel, one time for every ray. The accumulation can be compared with filling a histogram. For a series of increments at random histogram buckets, the design challenge of such an accumulator is known as "scattered add" [141, 142]. It exhibits the following problems for dual-port BRAM:

- Limited resources for adders: With 1024^2 detector pixels, resources on the DFE do not allow to consume one fixed-point accumulator for every detector pixels. Hence, only one fixed-point accumulator is used and the pixel values are stored in BRAM and retrieved from it during accumulation.
- Memory conflicts: An increment consists of a read access, followed by a write access to the same address for the updated value. BRAM has a nonzero latency and the value could have been updated in the clock cycle before, but the update could have not reached the BRAM yet. This read-after-write conflict can lead to lost updates.
- Clock speed: For an entire megabyte of BRAM, the read and write latency increases and requires either more clock cycles or a slowed-down clock frequency.

Fortunately, we can avoid all three problems after examining the access pattern on the virtual detector. The three-dimensional volume is traversed by incrementing the x index first to address the current voxel. When the x index wraps, the y index is incremented by one, until a voxel layer was visited. Only then is the z index incremented.

When only the x index of the current voxel in the volume is increased, the addresses of the four shadowed virtual detector pixels change monotonic because voxel and pixel positions are connected by an affine projection. In this case, the location of the detector value that has to be incremented by the shadow must be either the same as in the previous cycle or the location advances and will not return to a previous one. It is therefore sufficient to only handle the case that a pixel is incremented consecutively. Otherwise, we can load from the BRAM, increase, and write back without having to mind storage conflicts.

When the x index of the current pixel wraps, the same insight does not immediately hold true and a storage conflict could happen, since the wrap caused the pixel location to leap and to not move monotonically. The situation is sketched in Figure 4.27a for a very small volume with an edge length of only 14 voxels. The x index wraps between time steps 13 and 14. The wrap causes not the same, but the previous odd pixel to be hit again. The odd pixel that was hit in time step 6, the last time, is hit again in time step 14, disturbing the monotonic pixel access pattern. Also, at time step 21, the same pixel is increased that was already increased in time step 13.

For the previous pixel to be hit again in the worst case, the rays must be almost parallel to the direction of the x index. We therefore can at least afford a latency of $21 - 13 = 8$ clock cycles for the BRAM to store an update safely.

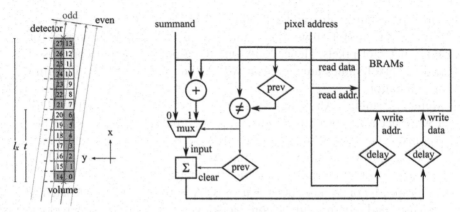

(a) Detector access pattern   (b) BRAM accumulator design before scheduling

Figure 4.27   The projection accumulator. The individual shadows of the voxels are accumulated on the virtual detector during forward projection. The access pattern on the detector is sufficiently monotonic to allow the latency of the BRAM to be mostly ignored, avoiding the general histogram problem (a). The hardware design only needs to check for sequences of accumulations to the same BRAM address. The multiplexers for the line transfers and the logic to ignore accumulations with a zero summand were left out for clarity (b).

The allowed BRAM latency increases for volumes with larger edge length. The time t between repeated hits is given in Equation (4.68) and was derived from Figure 4.27a. All units are multiples of voxels for l_x and clock cycles for t.

$$t = \left\lceil l_x - \sqrt{\frac{l_x^2}{4} - 1} \right\rceil \tag{4.67}$$

$$\simeq \frac{l_x}{2} \tag{4.68}$$

The time between repeated, nonmonotonic hits t is about half the edge length in x direction. In practice, the shortest edge of the volume is at least 200 voxels wide; hence, the hardware has a comfortable constraint of 100 clock cycles for a round trip between write to and read from BRAM. The implementation requires only 13 clock cycles. We are left with a small modification of the projection accumulator to ignore increments by zero, because the same odd ray that intersects the current voxel at time step 21 for the first time with a nonzero intersection length will already be sent to the accumulator with a zero intersection length since time step 18.

When the y index wraps, a voxel layer was completed and the forward projection FSM switches into a transfer state and projection pauses. When the forward projection continues again, all updates have been safely written to the BRAM.

The logic for the middle layer of the projection accumulator before static scheduling is shown in Figure 4.27b. The pixel address is compared with the previous one ($\neq$ node), and the value at the address is retrieved from the BRAM. If the address has changed, the value from the BRAM is added with the summand (i.e., the shadow of the current voxel) and loaded into the previously zeroed fixed-point accumulator ($\sum$). If the address has stayed the same, only the summand is sent to the fixed-point accumulator. The result of the accumulation is sent to the BRAM. The delay is required to allow the loop in the design to be scheduled and depends on the latency of the BRAM and the other logic in the loop's path. In the final hardware design, its latency is 13 clock cycles.

The logic for the BRAM accumulator contains logic that was not shown for increased clarity. It contains additional multiplexers to allow the readout and zeroing of detector lines for line transfers to the back projection kernel. Also, the logic that ignores summands that are equal to zero is not shown. This design is instantiated four times per voxel, one time for each ray.

The memory is assembled from many individual BRAM blocks on the lowest level of the projection accumulator. Xilinx ISE allows us to combine multiple BRAM slices on the FPGA into one big BRAM. This technique is helpful, but not sufficient for memory blocks that approach a quarter of a megabyte. The interconnects between the slices decrease the maximum achievable clock frequency drastically. The memory was therefore assembled manually in MaxJ, and layers of pipeline registers were added to limit path lengths and fanout. The optimal size for a BRAM block in terms of clock frequency was found to be close to the hardware size of a BRAM at 4 kB with two stages of extra pipeline registers before and after its ports.

The final accumulator design is able to accumulate four values per clock cycle and integrates seamlessly with the remaining pipeline stages of the forward projection kernel. It is controlled by the FSM of the kernel and notifies the FSM when the last detector line of an iteration was transferred to the back projection kernel.

4.2.4.5 Residues storage
The backward kernel receives the virtual projection from the forward projection kernel as well as the real projection and the total intersecting length of each ray from the host. After the virtual and the real projection were compared, the resulting residues need to be stored for lookup during back projection. Much

like the projection accumulator, only a band of pixel lines is kept in memory. Also, the storage is split into multiple parts to allow simultaneous read access for four residues per clock cycle and voxel.

The logic for the residue storage shares most components from the projection accumulator. For example, the storage part is implemented the same way with the same memory size by combining multiple BRAMs with extra pipeline registers. Only its core, the actual accumulator logic, is missing, since a residue is only written once during the transfer cycles of the back projection FSM. If a line of residues was not accessed during the back projection of the last layer of voxels, it is marked for removal and will be substituted by a new line during the following transfer.

Together with the residues storage, the FSM ensures that all pixel values from the forward projection kernel are consumed. Otherwise, the entire design could block when the output of the forward projection kernel stalls.

4.2.4.6 Multi-piping
A first implementation of the hardware processed one voxel per clock cycle. The design usage of this design was less than one third for all types of FPGA resources. To improve the efficiency, the pipelines in the kernels were multiplied. Flynn's taxonomy [29] of the architecture is thereby converted from a one-pipeline design with *multiple instruction streams, single data stream* toward a multi-piped design with *multiple instruction streams, multiple data streams.*

The resource usage of the single-piped design allowed us to estimate that the Xilinx Virtex-6 SX475T FPGA would be able to support a triple pipelining. Many components could just be multiplied by three, such as the ray–voxel intersection, or the width of the DRAM interface. Some components stayed the same, with modified parameters, such as the FSMs in both kernels.

An increased effort was needed for the projection accumulator and the residue storage. Both act as nontrivial reductions in the data stream. Voxels that were hit by the same ray at forward projection can now hit up to three voxels during the same clock cycle, and their shadow at the ray's endpoint on the detector must be added simultaneously instead of consecutively. Also, three voxels in a line can now shadow up to 2×4 pixels instead of 2×2 pixels (Figure 4.28).

The modification of the design therefore consisted of two parts. First, the partitioning of the memory for the projection accumulator and for the residue storage was changed. Before, it consisted of four parts, where each part was chosen whether the row and line address is odd or even. Now, the modulo

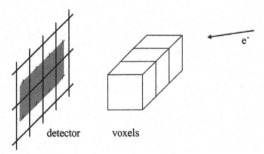

Figure 4.28 Shadow of a three voxels on the detector. Three voxels in a line can shadow up to 2 × 4 voxels. To simultaneously update all affected pixels, the BRAM of the projection accumulator was partitioned into 8 parts. The part is selected by calculating (x mod 4, y mod 2) in detector coordinates with bit slicing; the remaining bits are used as the address within the parts.

by four is calculated for the line address and selects the correct memory part together with the modulo by two of the row address. This allows us to perform eight simultaneous read and write accesses to the BRAM memory with the given access pattern. On the hardware level, we compute $(x_{\text{BRAM}}, y_{\text{BRAM}}) = (x \bmod 4, y \bmod 2)$ by slicing the last bits and use the remaining bits as the address within the parts.

The projection accumulator expects the shadows of the voxels to be already sorted and reduced with respect to the 2 × 4 memory partitioning. If a ray hits more than one of the three voxels, the shadows for the detector pixel at the endpoint have to added before. Pixels with the same pair of $(x_{\text{BRAM}}, y_{\text{BRAM}})$ are sorted in a short sorting network, and shadows are added if more than one of the three voxels is hit by the same ray on the way. The same sorting logic without the adders is also used for the residue storage in the back projection kernel.

The multi-piping logic requires the x side of the volume to be a multiple of three. In the worst case and for a volume with 1024^3 voxels, the volume must be padded with extra voxels that increase the voxel count by about 2‰. The overall speed, however, is still accelerated by about a factor of three.

4.2.5 Results

In this section, the results for the application for electron tomography are presented. After forward projection, comparison, and backward projection were all implemented on the DFE, we examine the application in terms of accuracy, speed, and resource usage.

4.2.5.1 Accuracy

The precision of the fixed-point data encodings was derived theoretically for most types from the resolution of the CCD camera pixels and the dimensions of the volume and its voxels (Table 4.4). The accumulators for the forward projection, the comparator for the residues, and the residue storage therefore compute exactly without error.

The matrix multiplications for the affine transformations between the coordinate systems of the volume and the microscope inevitably cause a loss in precision. Also, the ray–voxel intersection is susceptible to a noncontinuous change of the intersection length when a ray is parallel to one of the edges of a voxel. An arbitrarily small error in the voxel position can then make a difference between a full hit and a miss.

Figure 4.29 shows the error of the logic for ray–box intersection. The endpoint of the ray as well as the position of the voxel was randomly chosen 10,000 times within the possible boundaries of the application. Then, the ray direction was taken from a random distribution such that the voxel was hit in

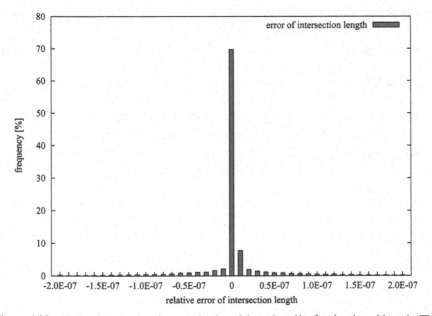

Figure 4.29   Error of the ray–box intersection length introduced by fixed-point arithmetic. The relative error of the ray–voxel intersection has a long tail. Its average is $\sigma_{\text{ray}-\text{voxel}} = 1.6 \cdot 10^{-7}$, which is two orders of magnitude smaller than the relative error introduced by the camera ($\sigma_{\text{CCD}} \geq 2^{-16} = 1.5 \cdot 10^{-5}$). The accumulators and the comparison logic for the residues calculate exactly.

about 50% of all cases by the single ray, an upper limit to the observed hit rate in the application with four rays per voxel. The error was determined by calculating the difference between the ray–voxel intersection length in Java with double precision and the bit-accurate simulation with the MaxCompiler library.

For a voxel with an edge length equal one, the standard deviation of the error is $\sigma_{ray-voxel} = 1.6 \times 10^{-7}$ with a bias of $\mu_{ray-voxel} = -3.8 \times 10^{-10}$. All errors stayed within $\pm 1.1 \times 10^{-6}$. The average error $\sigma_{ray-voxel}$ is far smaller than the error introduced by the CCD camera. Here, the relative error of a 16-bit integer is at least $\sigma_{CCD} \geq 2^{-16} = 1.5 \cdot 10^{-5}$. The relative error from the CCD is therefore at least two orders of magnitude larger than the error introduced by fixed-point arithmetic. A shadow is later calculated in the forward projection by multiplying the voxel density with the intersection length, and the relative errors of the factors are added quadratically. This allows us to ignore $\sigma_{ray-voxel}$ for the following (exact) projection accumulation and comparison with the real projection. The same applies for the back projection.

4.2.5.2 Throughput

For both the forward and the back projection kernel, a clock frequency of 170 MHz was achieved. The remaining parts of the design, such as the manager, the DRAM controllers, and the PCIe controller, were set to run at a slower clock frequency of 120 MHz. The different clock speeds allowed to reach timing closure for all parts of the design with a maximum speed for the kernels. The data paths of the slower parts were widened by a factor of two to compensate for the decreased clock frequency. The design was synthesized for the Max3 board with a Xilinx Virtex-6 SX475T FPGA [125].

The maximum throughput can be derived immediately from the parallelism within the kernels. Both kernels consume three pixels per clock cycle when they are not within a wait state or transfer data between the forward projection accumulator and the residue storage. The result for two different beam declination angles α is shown in Figure 4.30. The runtime of the application was measured as a function of the number of projections. For each projection count, the runtime was measured 10 times and the mean and the variance were plotted. A projection consisted of 1024×1024 values from the detector pixels, and the volume was partitioned in 1024^3 voxels.

The throughput is then derived from the slope of the trend line. With a clock frequency of 170 MHz, the maximum throughput would be 510 megavoxels/s in theory. For a declination angle of $\alpha_1 = 1°$, a throughput of

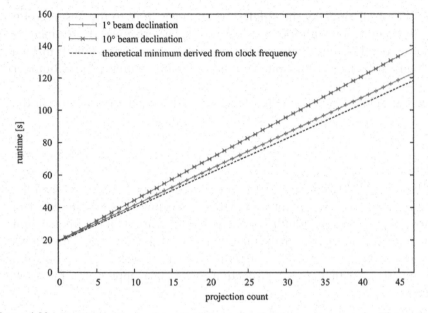

Figure 4.30   Runtime of the hardware implementation for electron tomography. The runtime is shown for the reconstruction of a volume with $1k^3$ voxels from projections with $1k^2$ pixels. The measured throughput is 490 megavoxels/s for a beam declination of $\alpha_1 = 1°$, which is close to the theoretical maximum of 510 megavoxels/s. The throughput is 425 megavoxels/s for an extreme declination angle of $\alpha_2 = 1°$ [125].

490 megavoxels/s was achieved, which translates to an efficiency of 96% when compared to the theoretical maximum. For $\alpha_2 = 1°$, the throughput decreases to 425 megavoxels/s, or 83% of the theoretical maximum. The variance of the runtime is low, smaller than 0.5% of the mean for each projection count.

The decrease in throughput correlates with an increased number of projection lines that have to be buffered at the start of a projection and before the back projection kernel can begin to update voxel densities. Similarly, projector lines that have not been transferred yet will block the forward projection kernel at the end of a projection until they are consumed by the back projection kernel. The amount of virtual detector lines where both kernels overlap is only 4 lines for α_1, but 250 lines for α_2 as the declination angle.

The combined (constant) setup and cleanup time is 19 s in both cases. The measurement did not include the time needed at the end of a reconstruction to write the results to a hard disk. The time needed to store a projection with a fixed number of voxels would add to the constant setup and cleanup time and introduce additional jitter.

4.2.5.3 Resource usage

The hardware design was implemented on the MAX3 board, which was big enough to hold all components of the algorithm with three parallel voxel pipelines. Table 4.5 lists the resources which were consumed on the embedded Xilinx Virtex-6 SX475T FPGA. The row for ray-tracing describes the resources required for the affine transformations and calculation for the ray–voxel intersection. The logic is present in both the forward and back projection kernel. It consumes most of the LUTs, FFs, and DSPs. The projection accumulator in the forward projection kernel and the residue storage in the back projection kernel mainly consist of BRAM, where the active part of the projection is buffered.

The kernels for forward and back projections consume about the same amount of resources within a 10% margin. The logic needed for the synthesized DRAM and PCIe controllers as well as the synchronization manager is expensive in terms of resource usage and consumes more lookup tables than both kernels combined. Apart from the DSP usage, the FPGA is occupied by about 50%. The number of available DSPs constitutes the limiting resource of the design. With more DSPs on the chip, the design could be extended to more than three voxel pipelines.

The numbers shown in Table 4.5 differ slightly from the previous publication [125] due to bug-fixes and optimizations that were introduced after submission. Most resources were saved when the control logic for forward and back projection was moved from a convoluted dataflow description toward a simpler and clearer description with state machines. An initial implementation with FSM did not reach 170 MHz for the kernels and was only reached again

Table 4.5 Resource usage on a Xilinx Virtex-6 SX475T FPGA (MAX3 DFE) for electron tomography. Ray-tracing consumes the DSPs, and the BRAMs are used for projection and residue storage. The table contains bug-fixed and optimizations introduced past the previous publication [125]

Component	LUTs	FFs	BRAMs	DSPs
Ray-tracing (per kernel)	22,220	66,939	0	720
Projection accumulator	5038	15,994	192	0
Residue storage and update	3985	13,728	192	0
Forward projection kernel	28,345	85,461	192	720
Back projection kernel	30,789	89,295	192	728
IO/DRAM/synchronization	72,936	90,942	181	0
Total resource usage	132,070	265,698	565	1448
Total resources available	297,600	595,200	1064	2,016
Total resource usage ratio	44.4%	44.6%	53.1%	71.8%

after the introduction of extra pipeline stages. These stages were added at the interface between the FSMs and the pipeline parts they control.

The source code of the e-tomography design consists of 7752 lines of MaxJ for the hardware description and 1569 lines of C++ for the host code. From the MaxJ description, the MaxCompiler library generated 364,936 lines of VHDL code. The longest path in the pipelines has a length of 97 stages in the kernel for forward projection and 109 stages for back projection. It took the author approximately eight months to implement and test the design.

4.2.6 Discussion

The algorithm for electron tomography was successfully ported to reconfigurable hardware by re-organizing the access order of the data toward a linear pattern. The new dataflow then gave us the opportunity to spend the same amount of time for each voxel for calculating the ray–voxel intersection instead of pseudo-random sampling along a ray. This leads to a pipeline design where both kernels could be statically scheduled.

The results show that we maintain accuracy where we cannot compute with otherwise exact fixed-point operations. The error introduced at the calculation of the ray–voxel intersection is two orders of magnitude below the error caused by the microscopy setup and therefore negligible. When the setup changes, the design can be parameterized with more integer or fractional bits for the fixed-point data types, depending on the constraints of the microscope.

The speed of the design slightly depends on the beam declination angle α. For a small angle, that is, for an electron microscope with well-calibrated optics, it closely approaches the maximum throughput derived from the clock frequency of the kernels. For a larger declination angle α, the number of transfer cycles increases. These cycles are a necessity that pauses the computation of the projections and therefore slow down the overall throughput. About 1/8 of all cycles are transfer cycles for the upper limit of $\alpha = 10°$.

The most recent design on graphics cards by Wei Xu et al. reaches more than 1 gigavoxel/s for some problem sizes [136]. Compared to this, the dataflow design is slower. However, the graphics card design relies on $\alpha = 0°$ and can therefore compute 3D reconstructions from a stack of 2D reconstructions. The required support for nonzero declination angles has complicated the design for 3D ray–box intersection and projection handling.

The graphics card that we received as a starting point from the research group of Achilleas Frangakis was developed on two NVIDIA Tesla C1060 working in parallel. These can trace rays from any direction and processes

196 megavoxels per second. The dataflow design therefore runs faster and accelerates reconstruction by a factor of 2.5 for a small, but common declination angle. Compared to a single graphics card, the application is accelerated by a factor of 5.

The dataflow implementation can be parallelized similar to the GPU implementation by cutting the voxel volume in multiple parts orthogonally to the rotation axis, similar to the graphics card implementation, The graphics card consumes about twice the space of a DFE. For the same rack space, the dataflow implementation is therefore faster by a factor of 10.

Due to the entirely different architectures, the reason for the acceleration must stay unclear. The dataflow design benefits from the linear access pattern on the volume storage in DRAM, and the continuous operation on numbers encoded as fixed point instead of floating point.

5

Conclusion

In the introduction of this book, three requirements were listed for a successful acceleration of applications from biomedical image processing and reconstruction with dataflow on FPGAs (see page 2). First, a program must be effectively ported from a description in an imperative language. Second, development in a dataflow language must be efficient. Finally, the program needs to run faster on the new hardware to justify the effort. The results in the previous chapter now give us the opportunity to revise these for both applications that were chosen as benchmarks: localization microscopy and electron tomography.

5.1 Portability

Both applications had their core algorithms written in the style of imperative control flow. It was not initially apparent whether the algorithms could be rewritten to follow a dataflow description. Before, smaller algorithms for image processing were implemented successfully on reconfigurable hardware, such as pixel functions and convolutions on raster images, but have stayed below the threshold of a fully ported, scientific, high-performance application. Larger applications, such as SART for computer tomography, were only partially implemented and have become suitable for reconfigurable hardware only recently after chip sizes have sufficiently increased. In this book, the image processing applications for localization microscopy and electron tomography (SART) were implemented with all compute kernels on reconfigurable hardware for the first time.

The port of both applications was conducted by focusing on the smallest unit of visual data, the pixel, or for 3D electron tomography, the voxel. The analysis of this dataflow then allowed the exploration of the solution space compatible with a pipeline architecture, and the selection of a solution that adheres to the constraints of the FPGA hardware and the periphery on its circuit board. For an efficient mapping, a solution that consists of a system of long and statically scheduled pipelines was found. These pipelines also contained few branches that would be seldom used for meaningful result.

Other applications that process rasterized imagery can likely be ported, too. The implementation of the algorithms for localization microscopy and electron tomography shows an inroad into the problem domain. The general approach was described in Chapter 3, "Acceleration of Imperative Code with Dataflow Computing." The chapter described both the example applications and showed the maintenance of the same functionality, including the accuracy of the computation.

5.2 High-Level Development

The increased size of the code base made the development of an implementation for a single developer unlikely with commonly used low-level description languages, such as VHDL and Verilog. The integration of a PCIe interface, for example, is estimated to already take half a year or more for a Ph.D. student. The time needed for implementation of the sample applications was only three months for localization microscopy and eight months for electron microscopy. The remaining time could be spent for high-level design and evaluation for accuracy, throughput, and resource usage.

The chosen MaxJ compiler library from Maxeler Technologies enabled the delivery of both sample applications within three years of this Ph.D. study. Contrary to other high-level languages that aim to translate imperative control flow descriptions to hardware, MaxJ follows the dataflow model to abstract from the low-level details of the hardware. This means that the hard work of translating control flow to dataflow remains at the programmer, and MaxJ can be understood more as a set of macros that generate hardware pipelines than a programming language where the underlying hardware can be of little interest to the developer. It is, however, the opinion of the author that the approach of MaxJ makes the resource usage much more predictable from the source code when compared to control flow abstractions.

The ratio of handwritten lines of code versus the amount of VHDL generated by the MaxCompiler library approaches 50 for the electron tomography application and gives a hint about the efficiency that can be gained through high-level tools. A comparison of the resource usage of handwritten VHDL and handwritten MaxJ was carried out by group member Heiko Engel [143]. It shows that the resource usage of MaxJ has a different signature in terms of lookup tables, flip-flops, and DSPs, but keeps the overall resource usage of the FPGA close to the VHDL implementation. Combining these results, MaxJ can be recommended as a tool to improve the efficiency of a hardware programmer for applications where hardware from Maxeler is an option.

5.3 Acceleration

The solution space was large enough to select for details such as the access pattern for the sample applications. The feature extraction kernel for localization microscopy allowed linear access of the signal in the region of interest, and the application for electron tomography could be rewritten to also access the 3D volume linearly voxel by voxel. The pipelines could then be scheduled statically to process one data item per clock cycle in general. The following benefits allow a dataflow pipeline to run faster than a CPU program despite slower clock frequencies and the infrastructure overhead on the chip needed for reconfiguration.

- **Register transfers:** General-purpose hardware has only a limited number of registers, and a compute core performs transfers between these registers one at a time. Techniques such as hyper-threading and multi-cores cannot ease this constraint to the same extent as a statically scheduled pipeline that performs hundreds of register transfers between pipeline stages at every clock cycle. For image processing, the pipeline is a natural way of processing pixel or voxel values for a wide range of algorithms, ranging from convolutions with a stencil to feature extraction and image reconstruction through back projections.
- **Control logic:** An application that can be written as a statically scheduled pipeline follows the dataflow and contains few control logic. The space required for the cache hierarchy, out-of-order execution, branch prediction, and virtual memory on general-purpose hardware on a CPU is freed and compensates for the overhead required for reconfiguration on a DFE. For image processing, both the example applications benefited from pipelines that consume one pixel or voxel per clock cycles and transport these through the stages before the corresponding intermediate values get accumulated or otherwise reduced for the end result. The dataflow pipeline can be seen as a member of the MapReduce family of algorithms.
- **Custom data operations:** FPGAs support numeric data types of custom range and precision. General-purpose hardware can only support a limited number of types, such as integers with a width that must be a power of two between 8 bits and 128 bits. The arithmetic logic in a CPU

must then follow a design that produces correct results for the worst case, independent of the error semantics. On the DFE, we are free to choose any encoding between these formats, depending on the intrinsic error of the input data. Image processing is especially well-suited for custom data operations because image data are recorded as integer pixel values. The following pipeline stages can then often be implemented with custom fixed-point hardware. Fixed-point operators consume less resources than floating-point operators on an FPGA. Again, the hereby saved resources become available for computing.

- **Larger FPGAs:** Finally, the availability of larger chips pushes the development of image processing applications on DFEs. The applications for electron tomography were only partially implemented on reconfigurable hardware before. An increase in chip resources that outpaces Moore's law allowed the transition of small applications that perform few convolutions to full-sized implementations for high-performance computing. While this enables portability in the first place, the increase in chip size makes the previous points to come into effect.

The results are execution times that are only fractions of what they were before. The analysis for localization microscopy was accelerated by a factor of 18,500, where a factor of 185 was due to hardware acceleration and the remainder due to changes in the algorithm. With equal rack space, hardware acceleration with dataflow engines achieved a factor of 62. For electron tomography, the application was accelerated by a factor of five compared to an implementation that was heavily optimized to run on a NVIDIA Tesla C1060 graphics card, or by a factor of ten when compared in terms of rack space.

It has already been shown that dataflow computing on reconfigurable hardware accelerates a wide range of applications by orders of magnitude compared to execution in CPUs [144]. With this work, it was shown that biomedical image processing and reconstruction can benefit from this approach in the same range.

5.4 Outlook

The acceleration of analysis for localization microscopy is expected to further advance optical light microscopy. The resolution of future CCD sensors with single-photon efficiency will increase further with time, and fluorophores that switch states at higher frequencies are in development. This will lead to bigger image frames and higher frame rates from the camera. The developed dataflow accelerator will keep up data processing speeds with the increased data rates and gives way to the next generation of localization microscopy. Images of life cells could then be resolved in real time with subsecond latencies, a timescale where many biological processes take place and localization microscopy becomes feasible to monitor in vivo manipulation.

For electron tomography, the acceleration factor of five is about the same factor that the DFE was more expensive than the graphics card. Further research and the observation of market prices will decide which hardware to use for future experiments. Here, better sensors with higher resolutions will allow bigger sample sizes, and the DFE is in a unique position to keep the entire 3D density distribution that will grow with $O\left(\beta\right)$ in its onboard memory.

FPGAs have evolved from small chips previously only used for "glue logic" to ASIC replacements and accelerators of fully featured applications. DFEs based on FPGAs are still less known and have remained a tool for special domain experts. With the acceleration potential shown in this book for applications that would have only fit on a CPU system few years ago, and the advent of high-level dataflow description languages that increase productivity while maintaining execution speeds, the distribution of dataflow accelerators is expected to further increase in popularity in the scientific community concerned with high-performance computing.

References

[1] James, R. M. (1982). Data-flow computing: the val language. *ACM Trans. Program. Lang. Syst.* 4, 44–82.

[2] Hurson, A. R., and Kavi, K. M. (2007). *Dataflow Computers: Their History and Future.* (New York: Wiley).

[3] Sung-Eun, C., and Lewis, E. C. (2000). A study of common pitfalls in simple multithreaded programs. *SIGCSE Bull.* 32, 325–329.

[4] Dennis, J. B., and Misunas, D. P. (1974). A preliminary architecture for a basic data-flow processor. *SIGARCH Comput. Archit. News* 3, 126–132, December.

[5] Dekker, S. T., Jonker, P. P., and Groen. F. C. A. (1987). "Distance transforms with data flow techniques," in *ASST '87 6. Aachener Symposium für Signaltheorie,* ed. D. Meyer-Ebrecht. Vol. 153, *Informatik-Fachberichte* (Berlin Heidelberg, Springer), 269–272.

[6] Vedder, R., and Finn, D. (1985). The hughes data flow multiprocessor: architecture for efficient signal and data processing. *SIGARCH Comput. Archit. News* 13, 324–332.

[7] Watson, I., and Gurd, J. (1899). "A prototype data flow computer with token labelling," in *International Workshop on Managing Requirements Knowledge* (Washington, DC: IEEE Computer Society), 623.

[8] Arvind and Nikhil, R. S. (1990). Executing a program on the MIT tagged-token dataflow architecture. *IEEE Trans. Comput.* 39, 300–318.

[9] Lakshmi Narasimhan, V. (1989). *Design and implementation of a dynamic dataflow array processor system (PATTSY).* Ph.D. thesis, The University of Queensland.

[10] McGraw, J. R. (1982) The val language: description and analysis. *ACM Trans. Program. Lang. Syst.* 4, 44–82.

[11] McGraw, J., Skedzielewski, S., Allan, S., Grit, D., Oldehoeft, R., Glauert, J., Dobes, I., and Hohensee, P. (1983). Sisal: streams and iteration in a single-assignment language. Language reference manual, version 1.1. Technical report, Lawrence Livermore National Lab., CA, USA.

[12] Budiu, M. (2003). *Spatial Computation.* Ph.D. thesis, School of Computer Science, Carnegie Mellon University.

[13] Jähne, B. (2014) *Digital Image Processing, Chapter 4 Neighborhood Operations* (Berlin: Springer), 105–134.

[14] Gonzales, R. C., and Woods, R. E. (2008). *Digital Image Processing, Chapter 3 Intensity Transformations and Spatial Filtering* (New Jersey: Pearson Education), 104–198.

[15] Bobda, C. (2007). *Introduction to Reconfigurable Computing.* Springer, Berlin.

[16] *Virtex-6 FPGA Configurable Logic Block—User Guide,* February 2012. Xilinx Inc., San Jose, CA.

[17] *Spartan and Spartan-XL FPGA Families Data Sheet,* June 2008. Xilinx Inc., San Jose, CA.

[18] Kuon, I. C. (2008). *Measuring and Navigating the Gap between FPGAs and ASICs.* Ph.D. thesis, University of Toronto.

[19] *Virtex-6 FPGA Memory Ressources—User Guide,* September 2013. Xilinx Inc., San Jose, CA.

[20] IEEE standard VHDL language reference manual. *IEEE Std 1076-1987,* 1988.

[21] Chapman, K. *Get your Priorities Right—Make your Design Up to 50% Smaller,* October 2007. Xilinx Inc., San Jose, CA.

[22] *Xilinx Synthesis Technology (XST)—User Guide,* November 2009. Xilinx Inc., San Jose, CA.

[23] Grüll, F. (2009). *The Parallel Object Language—Development and Implementation of an Object Oriented Language for Partial Dynamic Reconfiguration on FPGAs.* Diploma Thesis, University of Heidelberg.

[24] Esmaeilzadeh, H., Blem, E., St. Amant, R., Sankaralingam, K., and Burger. D. (2011). "Dark silicon and the end of multicore scaling," in *2011 38th Annual International Symposium on Computer Architecture (ISCA),* 365–376, June.

[25] Hennessy, J. L., and Patterson, D. A. (2012). *Computer Architecture: a quantitative approach, Chapter 1.5 Trends in Power and Energy in Integrated Circuits* (Pennsylvania: Elsevier), 21–26.

[26] Rupley, J. (2012). "AMD's "Jaguar": A next generation low power x86 core", in *Hot Chips 24.*

[27] Patterson, D. A., and Sequin, C. H. (1981). "RISC I: A reduced instruction set VLSI computer," in *Proceedings of the 8th Annual*

Symposium on Computer Architecture, ISCA '81 (Los Alamitos, CA: IEEE Computer Society Press), 443–457.

[28] Keynote, J. D. (2009). "Numbers everyone should know," in *Large-Scale Distributed Systems and Middleware (LADIS).*

[29] Flynn, M. (1972). Some Computer Organizations and their Effectiveness. *IEEE Trans. Comput.* C-21, 948–960.

[30] Intel Corporation. (2013). *Intel 64 and IA-32 Architectures Software Developer's Manual,* vol. 2: Instruction Set Reference, A–Z.

[31] Flynn, M. J., and Rudd, K. W. (1996). Parallel architectures. *ACM Comput. Surv.* 28, 67–70.

[32] Amdahl, G. M. (1967). "Validity of the single processor approach to achieving large scale computing capabilities," in *Proceedings of the April 18–20, 1967, spring joint computer conference,* AFIPS '67 (Spring) (New York, NY: ACM), 483–485.

[33] Gustafson. J. L. (1988). Reevaluating Amdahl's law. *Commun. ACM.* 31, 532–533, May.

[34] Gajski, D. D., and Kuhn, R. H. (1983). Guest editors' introduction: new VLSI tools. *Computer* 16, 11–14.

[35] TIOBE Software BV. (2014). TIOBE programming community index. Available at: http://www.tiobe.com/

[36] Bacon, D. F., Rabbah R., and Shukla, S. (2013). FPGA programming for the masses. *Commun. ACM,* 56, 56–63.

[37] *Handel-C language reference manual,* 2007. Agility Design Solutions Inc. Available at: http://www.agilityds.com/agilityds.com

[38] Xilinx Inc. (2014). Vivado design suite user guide: high-level synthesis.

[39] Meeus, W., Van Beeck, K., Goedemé, T., Meel, J., and Stroobandt, D. (2012). An overview of today's high-level synthesis tools. *Des. Autom. Embed. Syst.* 16, 31–51.

[40] *ROCCC 2.0 User's Manual - Revision 0.6,* February 2011.

[41] Buyukkurt, B., Cortes, J., Villarreal, J., and Najjar, W. A. (2010). Impact of high-level transformations within the roccc framework. *ACM Trans. Archit. Code Optim.* 7, 17:1–17:36.

[42] Maxeler Technologies. (2011). *MaxCompiler white paper.* Available at: http://www.maxeler.com/content/software/

[43] The OpenSPL Consortium. (2013). *Openspl: Revealing the power of spatial computing, December 2013.* Available at: http://www.openspl. org

[44] Huang, S. S., Hormati, A., Bacon, D., and Rabbah, R. (2008). "Liquid metal: Object-oriented programming across the hardware/software boundary," in *ECOOP 2008,* Paphos, Cyprus.

[45] Eclipse foundation. The eclipse project. Available at: https://www.eclip se.org

[46] Silicon Software GmbH. *Silicon Software runtime 5 documentation.* Available at: http://www. siliconsoftware.de

[47] *Silicon Software VisualApplets Documentation.* (2010). Silicon Software, Mannheim.

[48] Kirchgessner, M. (2011) *FPGA-based hardware acceleration of localization microscopy.* Diploma thesis, University of Heidelberg.

[49] Knuth, D. E. (2005). *The Art of Computer Programming, vol. 1, Fundamental Algorithms, Chapter 2.2 Linear lists* (Boston: Addison-Wesley), 238–307.

[50] McCarthy, J. (1960). Recursive functions of symbolic expressions and their computation by machine, part I. *Commun. ACM,* 3, 184–195.

[51] Amamiya, M., Hasegawa, R., Nakamura, O., and Mikami, H. (1982). "A list-processing-oriented data flow machine architecture," in *Proceedings of the June 7–10, 1982, National Computer Conference,* AFIPS '82 (New York, NY: ACM), 143–151.

[52] Amamiya, M., Hasegawa, R., and Mikami, H. (1982). "List processing with a data flow machine," in *RIMS Symposium on Software Science and Engineering,* ed. E. Goto, K. Furukawa, R. Nakajima, I. Nakata, and A. Yonezawa, Vol. 147. *Lecture Notes in Computer Science* (Berlin: Springer), 165–190.

[53] Haskell Data.List library, version 4.6.0.1. Available at: http://www.hask ell.org/ghc/docs/7.6.2/html/libraries/base-4.6.0.1/Data-List.html

[54] American National Standards Institute, editor. (2003). *Programming languages—C++, Chapter 25. Algorithms Library,* 543–570. ISO/IEC 14882.

[55] Barrett, R., Berry, M., Chan, T. F., Demmel, J., Donato, J. M., Dongarra, J., Eijkhout, V., Pozo, R., Romine, C., and Van der Vorst, H. (1987). *Templates for the Solution of Linear Systems: Building Blocks for Iterative Methods,* Chapter Survey of sparse matrix storage formats (Philadelphia, PA: Society for Industrial and Applied Mathematics), 57–60.

[56] Knuth, D. E. (2005). *The Art of Computer Programming, Chapter 7.2.1.2 Generating all permutations, Algorithm L,* Vol. 4, fascicle 3 (Boston: Addison-Wesley), 5–6.

[57] Butler, J. T., and Sasao, T. (2012). "Hardware index to permutation converter," in *Parallel and Distributed Processing Symposium Workshops PhD Forum (IPDPSW), 2012 IEEE 26th International,* 431–436.

[58] Sipser, M. (1996). *Introduction to the Theory of Computation, Chapter 1. Regular languages.* Boston, MA: PWS Publishing.

[59] Knuth, D. E. (2005). *The Art of Computer Programming, vol. 3, Sorting and Searching, Chapter 5.3 Optimum sorting* (Boston: Addison-Wesley), 180–247.

[60] Koch, D., and Torresen, J. (2011). "FPGASort: a high performance sorting architecture exploiting run-time reconfiguration on FPGAs for large problem sorting," in *Proceedings of the 19th ACM/SIGDA international symposium on Field Programmable Gate Arrays,* FPGA '11. (New York, NY: ACM), 45–54.

[61] Chaudhuri, A. S., Cheung, P. Y. K., and Luk, W. (1997). "A reconfigurable data-localised array for morphological algorithms," in *Field-Programmable Logic and Applications,* vol. 1304 of *Lecture Notes in Computer Science.* (Berlin: Springer), 344–353.

[62] Graham, S. L., Kessler, P. B., and Mckusick, M. K. (1982). Gprof: A call graph execution profiler. *SIGPLAN Not.* 17(6), 120–126.

[63] VisualVM. *All-in-one Java troubleshooting tool.* Available at: http://visualvm.java.net

[64] Intel Corporation. (2013). *Intel 64 and IA-32 Architectures Software Developer's Manual,* vol. 3B: System programming guide, part 2, Chapter 18: Performance monitoring, 127–215.

[65] Linux profiling with performance counters. Available at: https://perf.wiki.kernel.org

[66] Mitchell, J. C. (2003). *Concepts in Programming Languages, Chapter Part 2, Procedures, Types, Memory Management, and Control* (Cambridge: Cambridge University Press), 93–228.

[67] The Go programming language. Available at: http://www.golang.org/ref

[68] Dijkstra, E. W. (1968). Letters to the editor: go to statement considered harmful. *Commun. ACM,* 11, 147–148.

[69] Böhm, C., and Jacopini, G. (1966). Flow diagrams, Turing machines and languages with only two formation rules. *Commun. ACM* 9, 366–371.

[70] Cytron, R., Ferrante, J., Rosen, B. K., Wegman, M. N., and Zadeck, F. K. (1991). Efficiently computing static single assignment form and the control dependence graph. *ACM Trans. Program. Lang. Syst.* 13(4), 451–490.

[71] Adé, M., Lauwereins, R., and Peperstraete, J. A. (1997). "Data memory minimisation for synchronous data flow graphs emulated on DSP-FPGA targets," in *Proceedings of the 34th annual Design Automation Conference,* DAC '97. (New York, NY: : ACM), 64–69.

[72] Warren, H. S. (2002). *Hacker's Delight.* Addison-Wesley, Boston).

[73] *Intel C++ Compiler XE 13.1 User and Reference Guide,* 2013.

[74] Sun, S., and Zambreno, J. (2009). "A floating-point accumulator for FPGA-based high performance computing applications," in *Proceedings of the International Conference on Field-Programmable Technology (FPT).*

[75] Maxeler Technologies. Acceleration tutorial, loops and pipelining. Version 2012.2.

[76] Wolfe, M. (1989). "More iteration space tiling," in *Proceedings of the 1989 ACM/IEEE conference on Supercomputing,* Supercomputing '89. (New York, NY: ACM), 655–664.

[77] Weaver, N., Markovskiy, Y., Patel, Y., and Wawrzynek, J. (2003). Post-placement C-slow retiming for the Xilinx Virtex FPGA. in *Proceedings of the 2003 ACM/SIGDA eleventh international symposium on Field programmable gate arrays,* FPGA '03 (New York, NY: : ACM), 185–194.

[78] Oracle. *Java Platform, Standard Edition 6, API Specification.* Available at: http://java.util.Random

[79] Knuth, D. E. (1969). *The Art of Computer Programming, vol. 2, Seminumerical Algorithms, Chapter 3.2.1 The Linear Congruential Method.* (Boston: Addison-Wesley) 9–10.

[80] Knuth, D. E. (1969). *The Art of Computer Programming, vol. 2, Seminumerical Algorithms, Chapter 4.3.3. How fast can we multiply?* (Boston: Addison-Wesley), 258–280.

[81] The Institute of Electrical and Inc. (2008). Electronics Engineers. IEEE standard for floating-point arithmetic 754–2008.

[82] *LogiCORE IP Floating-Point Operator v5.0,* March 2011. Xilinx Inc., San Jose, CA.

[83] Kingsbury, N. G., and Rayner, P. J. W. (1971). Digital filtering using logarithmic arithmetic. *Electron. Lett.* 7, 56–58.

[84] Lang, N. P. C. B. (2005). *Mathematische Methoden in der Physik, Chapter 21.2 Wahrscheinlichkeitsrechnung und Statistik.* (Pennsylvania: Elsevier), 629.

[85] Gowen, R. A., and Smith, A. (2003). Square root data compression. *Rev. Sci. Instrum.* 74, 3853–3861.

[86] Savakis, A., and Piorun, M. (2002). Benchmarking and hardware implementation of JPEG-LS. in *2002 International Conference on Image Processing. Proceedings,* Vol. 2, II 949–II 952.

[87] Lehtoranta, I., Salminen, E., Kulmala, A., Hannikainen, M., and Hamalainen, T. D. (2005). A parallel MPEG-4 encoder for FPGA based multiprocessor SoC. in *International Conference on, Field Programmable Logic and Applications, 2005,* 380–385.

[88] Stephenson, M., Babb, J., and Amarasinghe, S. (2000). Bidwidth analysis with application to silicon compilation. *SIGPLAN Not.* 35, 108–120.

[89] Advanced Micro Devices. (2012). *AMD64 Architecture Programmer's Manual,* vol. 2: System Programming, Chapter 2: x86 and AMD64 Architecture Differences. September.

[90] JEDEC solid state technology association. (2012). *DDR3 SDRAM Standard,* Chapter 3.4.2.1: Functional Description—Burst Length, Type and Order. July.

[91] JEDEC solid state technology association. (2012). *DDR3 SDRAM Standard,* chapter 12.3: Electrical Characteristicx & AC Timing for DDR3-800 to DDR3-2133—Standard Speed Bins.

[92] Kaufmann, R., Müller, P., Hildenbrand, G., Hausmann, M., and Cremer, C. (2011). Analysis of her2/neu membrane protein clusters in different types of breast cancer cells using localization microscopy. *J. Microsc.* 242, 46–54.

[93] Abbe, E. (1882). The relation of aperture and power in the microscope. *J. Royal Microsc. Soc.* 2, 300–309.

[94] Broglie, L. (1970). The reinterpretation of wave mechanics. *Found. Phys.* 1, 5–15.

[95] Grüll, F., Kirchgessner, M., Kaufmann, R., Hausmann, M., and Kebschull, U. (2011). "Accelerating image analysis for localization microscopy with FPGAs," in *2011 International Conference on Field Programmable Logic and Applications (FPL),* 1–5.

[96] Hell, S. W., and Wichmann, J. (1994). Breaking the diffraction resolution limit by stimulated emission: stimulated-emission-depletion fluorescence microscopy. *Opt. Lett.* 19, 780–782, Jun.

[97] Rust, M. J., Bates, M., and Zhuang, X. (2006). Sub-diffraction-limit imaging by stochastic optical reconstruction microscopy (STORM). *Nat. Methods* 3, 793–796.

[98] Betzig, E., Patterson, G. H., Sougart, R., Lindwasser, O. W., Olenych, S., Bonifacino, J. S., Davidson, M. W., Lippincott-Schwartz, J., and

Hess, H. F. (2006). Imaging intracellular fluorescent proteins at nanometer resolution. *Sci. Express* 313, 1642–1645.

[99] Hess, S. T., Girirajan, T. P. K., and Mason. M. D. (2006). Ultrahigh resolution imaging by fluorescence photoactivation localization microscopy. *Biophy. J.* 91, 4258–4272.

[100] Airy, G. B. (1835). On the diffraction of an object-glass with circular aperture. *Trans. Cam. Philos. Soc.* 5, 283.

[101] Lang, N. P. C. B. (2005). *Mathematische Methoden in der Physik, Chapter 17.2 Die Besselsche Differenzialgleichung.* (Pennsylvania: Elsevier), 496.

[102] Andor Technology. *Sensitivity of CCD cameras, Key factors to consider.* White paper.

[103] Holden, S. J., Uphoff, S., and Kapanidis, A. N. (2011). DAOSTORM: an algorithm for high-density super-resolution microscopy. *Nat. Methods*, 8, 279.

[104] Cox, S., Rosten, E., Monypenny, J., Jovanovic-Talisman, T., Burnette, D. T., Lippincott-Schwartz, J., Jones, G. E., and Heintzmann, R. (2011). Bayesian localization microscopy reveals nanoscale podosome dynamics. *Nat. Methods* 9, 195–200.

[105] Lemmer, P., Gunkel, M., Baddeley, D., Kaufmann, R., Urich, A., Weiland, Y., Reymann, J., Müller, P., Hausmann, M., and Cremer, C. (2008). SPDM: Light microscopy with single-molecule resolution at the nanoscale. *Appl. Phys. B: Las. Opt.* 93, 1–12.

[106] Brandt, S. (1998). *Data Analysis, Chapter 9. The Method of Least Aquares.* Springer, Berlin.

[107] Press, W. H., Teukolsky, S. A., Vetterling, W. T., and Flannery, B. P. (2007). *Numerical Recipes—The Art of Scientific Computing, Chapter 15.5 Nonlinear Models.* (Cambridge: Cambridge University Press), 799–806.

[108] The MathWorks Inc. *MATLAB and Simulink R2013a documentation: Solve nonlinear curve-fitting (data-fitting) problems in least-squares sense.* Available at: http://www.mathworks.de/de/help/optim/ug/lsqcurv efit.html

[109] Brandt, S. (1998). *Data Analysis, Chapter 7. The Method of Maximum Likelihood.* Springer, Berlin.

[110] Feng, Y., Goree, J., and Liu, B. (2007). Accurate particle position measurement from images. *Rev. Sci. Inst.* 78, 053704–053710.

[111] National Institutes of Health. *ImageJ, image processing and analysis in Java.* Available at: http://rsb.info.nih.gov

[112] Thompson, R. E., Larson, D. R., and Webb, W. W. (2002). Precise nanometer localization analysis for individual fluorescent probes. *Biophys. J.* 82(5), 2775–2783.

[113] Andersson, S. B. (2007). "Precise localization of fluorescent probes without numerical fitting," in *4th IEEE International Symposium on Biomedical Imaging: From Nano to Macro, 2007. ISBI 2007*, 252–255.

[114] Bancroft, S. (1985). An algebraic solution of the GPS equations. *IEEE Trans. Aerosp. Electron. Syst.* AES-21, 56–59.

[115] Shen, Z., and Andersson, S. B. (2011). Bias and precision of the fluoroBancroft algorithm for single particle localization in fluorescence microscopy. *IEEE Trans. Signal Process.* 59, 4041–4046.

[116] Cheezum, M. K., Walker, W. F., and Guilford, W. H. (2001). Quantitative comparison of algorithms for tracking single fluorescent particles. *Biophy. J.* 81, 2378–2388.

[117] Quan, T., Li, P., Long, F., Zeng, S., Luo, Q., Hedde, P. N., Nienhaus, G. U., and Huang, Z.-L. (2010). Ultra-fast, high-precision image analysis for localization-based super resolution microscopy. *Opt. Express* 18, 11867–11876.

[118] *Dipimage and diplib*. Available at: http://www.diplib.org

[119] Brandt, S. (1998). *Data Analysis, Chapter 4. Computer Generated Random Numbers, The Monte Carlo Method*. Springer, Berlin.

[120] *The MathWorks Inc. MATLAB and Simulink R2013a documentation: Techniques for improving performance*. Available at: http://www.mathworks.de/de/help/matlab/matlab_prog/techniques-for-improving-performance.html

[121] *Qt online reference documentation*. Available at: http://doc.qt.nokia.com/

[122] Kubitscheck, U., Kückmann, O., Kues, T., and Peters, R. (2000). Imaging and tracking of single GFP molecules in solution. *Biophy. J.* 78, 2170–2179.

[123] Kaufmann, R., Piontek, J., Grüll, F., Kirchgessner, M., Rossa, J., Wolburg, H., Blasig, I. E., and Cremer, C. (2012). Visualization and quantitative analysis of reconstituted tight junctions using localization microscopy. *PLoS ONE*, 7, e31128.

[124] Sage, D., Kirshner, H., Pengo, T., Stuurman, N., Min, J., and Manley, S. (2013). "Localization microscopy challenge workshop," in *IEEE International Symposium on Biomedical Imaging 2013*.

[125] Grüll, F., Kunz, M., Hausmann, M., and Kebschull, U. (2012). "An implementation of 3D electron tomography on FPGAs," in *2012*

International Conference on Reconfigurable Computing and FPGAs (ReConFig), 1–5.

[126] Gordon, R., Bender, R., and Herman, G. T. (1970). Algebraic reconstruction techniques (ART) for three-dimensional electron microscopy and X-ray photography. *J. Theor. Biol.* 29, 471–481.

[127] Kaczmarz, S. (1937). Angenäherte Auflösung von Systemen linearer Gleichungen. *Bull. Intl. de l'Académie Polonaise des Sciences et des Lettres,* 35, 355–357.

[128] Saarbrücken Fraunhofer Institut für zerstörungsfreies Prüfen.

[129] Medalia, O., Weber, I., Frangakis, A. S., Nicastro, D., Gerisch, G., and Baumeister, W. (2002). Macromolecular architecture in eukaryotic cells visualized by cryoelectron tomography. *Science* 298, 1209–1213.

[130] Andersen, A. H., and Kak, A. C. (1984). Simultaneous algebraic reconstruction technique (SART): A superior implementation of the ART algorithm. *Ultrasonic Imag.* 6, 81–94.

[131] Bracewell, R. N. (1990). Numerical transforms. *Science* 248, 697–704.

[132] Herman, G. T. (2009). *Fundamentals of Computerized Tomography— Image Reconstruction from Projections, Chapter 8 Filtered Backprojection for Parallel Beams,* (Berlin: Springer), 135–157.

[133] Frank, J., Radermacher, M., Penczek, P., Zhu, J., Li, Y., Ladjadj, M., and Leith, A. (1996). SPIDER and WEB: Processing and visualization of images in 3D electron microscopy and related fields. *J. Struct. Biol.* 116, 190–199.

[134] Sorzano, C. O. S., Marabini, R., Velázquez-Muriel, J., Bilbao-Castro, J. R., Scheres, S. H. W., Carazo, J. M., and Pascual-Montano, A. (2004). XMIPP: a new generation of an open-source image processing package for electron microscopy. *J Struct. Biol.* 148, 194–204.

[135] Díez, D. C., Mueller, H., and Frangakis, A. S. (2006). Implementation and performance evaluation of reconstruction algorithms on graphics processors. *J. Struct. Biol.* 157(1), 288–295.

[136] Xu, W., Xu, F., Jones, M., Keszthelyi, B., Sedat, J., Agard, D., and Mueller, K. (2010). High-performance iterative electron tomography reconstruction with long-object compensation using graphics processing units (GPUs). *J. Struct. Biol.* 171, 142–153.

[137] Leeser, M., Coric, S., Miller, E., Yu, H., and Trepanier, M. (2005). Parallel-beam backprojection: An FPGA implementation optimized for medical imaging. *J. VLSI Signal Process. Syst.* 39, 295–311.

[138] Smits, B. (2005). Efficiency issues for ray tracing. in *ACM SIGGRAPH 2005 Courses,* SIGGRAPH '05. (New York, NY: ACM).

[139] Shirley, P., and Marschner, S. (2009). *Fundamentals of Computer Graphics, chapter 12.3.1 Bounding Boxes.* (Natick: A K Peters) 279–285.

[140] *Virtex-6 FPGA Families Product Table.* (2012). Xilinx Inc., San Jose, CA.

[141] Ahn, J. H., Erez, M., and Dally, W. J. (2005). "Scatter-add in data parallel architectures," in *11th International Symposium on High-Performance Computer Architecture, 2005. HPCA-11,* pp 132–142.

[142] Shahbahrami, A., Hur, J. Y., Juurlink, B., and Wong, S. (2008). "FPGA implementation of parallel histogram computation," in *2nd HiPEAC Workshop on Reconfigurable Computing, Göteborg, Sweden,* 63–72.

[143] Grüll, F., Engel, H., and Kebschull, U. (2013). "Signal processing in physics with high-level dataflow sysnthesis on FPGAs," in *ACM International Conference on Computing Frontiers.*

[144] Pell, O., and Averbukh, V. (2012). Maximum performance computing with dataflow engines. *Comput. Sci. Eng.* 144, 98–103.

[145] Mencer, O., Platzner, M., Morf, M., and Flynn, M. J. (2001). Object-oriented domain-specific compilers for programming FPGAs. *IEEE Trans. VLSI* 9, 205–210 (special issue on Reconfigurable Computing).

[146] ASC. (2006). A stream compiler for computing on FPGAs. *IEEE Trans. Comput. Aided Design Integr. Circ. Syst.* 9, 1603–1617.

Index

About the Authors

Frederik Grüll researches image processing with dataflow computing on FPGAs as a postdoc at Goethe University Frankfurt. He graduated in physics at the University of Heidelberg with computer science as a second subject and spent an exchange year at Imperial College London. During his final year project, he gained interest in the architecture of FPGAs and compiler development. After an internship for half a year at Maxeler Technologies, he started his PhD studies in the research group of Udo Kebschull and received his doctorate in computer science from Goethe University Frankfurt in 2015.

Udo Kebschull is a Professor at Goethe University Frankfurt, where he leads an FPGA research group that covers dataflow computing, radiation tolerance and detector read-out for high-energy physics. He graduated in computer science at the University of Karlsruhe and received his PhD from the University of Tübingen. In the following years, he was Professor for technical information science at the University of Applied Science Karlsruhe, the University of Leipzig and the University of Heidelberg before he moved to Goethe University Frankfurt in 2010. Here, he leads the research of infrastructure and computing systems and is head of the university's computing center.